Charles A. Witchell

The Evolution of Bird-Song

With Observations on the Influence of Heredity and Imitation

Charles A. Witchell

The Evolution of Bird-Song

With Observations on the Influence of Heredity and Imitation

ISBN/EAN: 9783337180805

Printed in Europe, USA, Canada, Australia, Japan

Cover: Foto ©ninafisch / pixelio.de

More available books at **www.hansebooks.com**

THE
EVOLUTION OF BIRD-SONG

WITH

OBSERVATIONS ON THE INFLUENCE OF
HEREDITY AND IMITATION

BY

CHARLES A. WITCHELL
AUTHOR OF 'THE FAUNA OF GLOUCESTERSHIRE'

LONDON
ADAM AND CHARLES BLACK
1896

PREFACE

I WISH to mention in this place my deep sense of obligation to those kind helpers who have given me notes on the subjects discussed in this book. I wish also to acknowledge my gratitude to the late Professor Harker, F.L.S., whose lamented death recently occurred. He, from the first, encouraged me to continue my observation of bird-song ; and, in 1890, he devoted a considerable amount of time to harmoniously arranging the results of my investigations.

When I state that Mr. J. E. Harting generously offered to look over the proof-sheets of this work (which offer was, of course, gladly accepted), the public will appreciate the extent of my indebtedness to an author and editor so accomplished. The first

32 pages had then been returned to the printer; but, in revising the remainder, I continually perceived the great advantages derived from Mr. Harting's perusal.

However novel or otherwise may be the theories stated in this book, I can at least claim that, so far as I am concerned, they are absolutely original, all of them having been committed to writing, though, in some instances, not under their present titles, before I consulted any person, or any book, in regard to them.

CHARLES A. WITCHELL.

LONDON, *April* 1896.

CONTENTS

INTRODUCTION . . . . . . PAGES
                                 I-II

CHAPTER I

THE ORIGIN OF THE VOICE

Darwin's opinion that the voice resulted from involuntary movements of muscles—Combat the chief occasion for such movements—Prevalence of a puff or hiss as a menace—Accidental cries of newts 12-21

CHAPTER II

ALARM-NOTES

Combat-cry serviceable as an alarm—Repetition of single alarm-cries, in terror—Theory of evolution of certain rattling cries by such repetitions—Discrimination of enemies evidenced by cries and deportment—Inherited knowledge of alarm-cries 22-32

CHAPTER III

THE INFLUENCE OF COMBAT

Rivalry and emulation—Among polygamous species, selection dependent on combat—The snapping of bill as a menace—Chaffinch's war-note—Singing during combat . . 33-40

CHAPTER IV

THE CALL-NOTE

PAGES

Erroneous descriptions of call-notes—Chaffinch's cry *pink* not a call-note—Distress-cries of young are of use as call-notes when the birds leave the nest—Influence of gregariousness and of fear—Absence of call-notes from the songs of extravagant singers: prevalence of them in songs of inferior singers   41-48

CHAPTER V

THE SIMPLEST SONGS

Repetition of call-notes in the breeding season—Construction of strains by this method—Examples—Call-notes concluding spring songs of skylark—Movements of wings during song —Repetition of same intervals of pitch   .   .   .   49-58

CHAPTER VI

NOTICEABLE INCIDENTS CONNECTED WITH BIRD-SONG

Songs are generally uttered by males: exceptions—Not until birds have attained full size: exceptions—Most frequently at morning and evening: influence of weather—Tendency to rise in pitch with vehemence—Only small birds properly sing—Singers arboreal birds generally—Effect of living amid foliage: on size, hearing, and voice—Accent in songs— Singers clad in sober hues—Development of the eyes in detecting danger—Necessity of leisure—Labours of parent-birds—Laborious and stealthy birds habitually poor in song —Flight in song: for purposes of display—Fluttering of wings a means of address—Ventriloquism—Singing in chorus   .   59-85

CONTENTS ix

CHAPTER VII

THE INFLUENCE OF HEREDITY IN THE PERPETUATION OF THE CRIES OF BIRDS

PAGES

Absolute inheritance of cries in plovers, common fowl, duck, swan, pheasant, etc.—Family cries—Cries of allied young more alike than those of allied adults—The *cahs* of the crows: heard in starling and jay—Similarity between starling and thrushes—Rattling alarm of mistle-thrush, modified in blackbird, ousel, song-thrush, fieldfare, and American robin—Similarity of occasions of utterance of alarms by blackbird and European robin—Similarities traced in call-notes of birds of the thrush family: also between their songs—Prevalence of imitation—The warblers—Particular cry common to nightingale, willow-warbler, and chiffchaff—Similar alarm-croaks of nightingale, sedge-warbler, and lesser whitethroat—Recapitulation—Similarities between notes of wagtails—Songs of tree-pipit and meadow-pipit described: similarity to that of skylark—Larks of three continents, with the same manner of song—Family resemblances in the buntings—Resemblance between call-notes of yellow bunting and greenfinch in flight—Notes of the finches—Canary has family traits—*Tell tell* cries of house-sparrow and greenfinch—Similar cries of young chaffinch and young house-sparrow—Various other orders mentioned 86-139

CHAPTER VIII

VARIATION IN BIRD-VOICES: ITS CAUSES AND EFFECTS

Prevalence of variation—Instances mentioned by authors—Variation in songs of blackbird, yellow-hammer, chaffinch, robin, house-sparrow, and cuckoo—Illustrated in the robin's alarm 140-158

CHAPTER IX

THE INFLUENCE OF IMITATION IN RELATION TO BIRD-SONG

PAGES

Imitation by dogs and other animals—The notes of some birds wholly perpetuated by imitation; of others wholly by inheritance—Observations of authors on the mimicry of birds—House-sparrow with lark's song, and other instances—Possible effects of imitation—Interesting similarities observable: between notes of birds and sounds produced by elements—the music of the streams—by insects, by quadrupeds, by birds—General observations—Chief subjects imitated by various birds—Thrushes, robins, skylarks, sedge-warblers—showed influence of arrival of summer migrants—Full records of songs of thrush, robin, skylark, starling, sedge-warbler, redstart, nightingale, marsh-warbler, wheatear, goldcrest, whitethroats, nuthatch, reed-bunting, stonechat, blackbird, chaffinch, and others 159-229

CHAPTER X

THE MUSIC OF BIRD-SONG

Repetition of intervals of pitch—Our diatonic scale was known 3000 years ago—Curious intervals sung by great titmouse, chaffinch, and robin—Curious crowing of fowls—Intervals in blackbirds' alarms—Music of the blackcap, mistle-thrush, and American robin 230-238

CONCLUSION 239-240

APPENDIX

Transcripts of music sung by blackbirds, thrushes, and skylarks 241-246

BIBLIOGRAPHY OF THE SUBJECT . . . 247, 248

INDEX 249-253

ERRATUM.

Page 103, paragraph 2 :
The cry of the young Nightingale is wrongly described here. It is not like that of the young Blackbird, but is a croak similar to the croak of the old Nightingales. I have already published an account of how this error was occasioned (*Zoologist*, No. 673, July, 1897, page 331).

CHARLES A. WITCHELL.

was aware of the imitativeness of many musical species when in captivity, and I was acquainted with the notes of most of the wild birds found in Gloucestershire. I listened to this nightingale, at the distance of a few yards, on eighteen out of the twenty-one evenings on which he sang. His nest

CHAPTER IX

THE INFLUENCE OF IMITATION IN RELATION TO BIRD-SONG

PAGES

Imitation by dogs and other animals—The notes of some birds wholly perpetuated by imitation ; of others wholly by inherit-

CONCLUSION 239-240

APPENDIX

Transcripts of music sung by blackbirds, thrushes, and skylarks 241-246

BIBLIOGRAPHY OF THE SUBJECT . . . 247, 248

INDEX 249-253

INTRODUCTION

THE idea of making a scientific investigation of the various features of bird-song first occurred to me in the year 1881, on the occasion of my listening to a nightingale near Stroud. I observed the frequent utterance of a little slurred whistle at the commencement of many of its songs—a whistle which I knew to be uttered by several of its congeners; and the songs themselves often seemed to include a repetition of the notes of birds of other genera with which I was acquainted. I had never read anything particular about the songs of birds, but was aware of the imitativeness of many musical species when in captivity, and I was acquainted with the notes of most of the wild birds found in Gloucestershire. I listened to this nightingale, at the distance of a few yards, on eighteen out of the twenty-one evenings on which he sang. His nest having been robbed, he afterwards uttered only one or two dismal notes at long intervals. I knew the song of this bird very well; and afterwards, in the same season, when listening to other nightingales, I soon learned the well-known fact that these birds, like many others, differ individually in their songs and times of singing. During the next few years I found no inducement to attempt any deliberate investigation of the songs of birds; but in 1885 the subject was again brought prominently before me. A thrush of very remarkable vocal powers had its nest in our garden, and sang every day in a high acacia-tree not far from the house. At that time I was much at home nursing a sick relation, and in these circumstances the loud clear song of the thrush was particularly noticeable. Towards the end of May we observed that on every successive morning the bird began to sing about a minute and a half earlier, so that we could predict almost to a minute the time when his first notes would be heard. His extraordinary endurance led me to record his periods of song, with the astounding result that these on one day amounted to an aggregate of no less than sixteen hours, thus allowing only three hours of silence during the whole time of song, which extended from 2.45

A.M. to 9.30 P.M. This thrush was the first wild one which I heard imitate. Like other persons fond of the country, I had listened to hundreds, perhaps thousands, of thrushes, and yet had never noticed their mimicry. The bird in question exactly reproduced the "call" of the partridge—I mean the cry heard so often on winter evenings—and this was the sound which first directed my attention critically to the character of the song. The thrush also repeated many cries then well known to me; for instance, the most prominent notes of the house-sparrow, greenfinch, and chaffinch; the loud, prolonged call of the blue titmouse; and notes of the wagtail, brown wren, and some other birds. The invalid, a brother who had been my frequent companion in the country, fully endorsed my opinion of certain notes of the singer; indeed, he was the first to identify its imitation of the brown linnet.

Having found mimicry in the thrush, I listened for it in the songs of other birds. In some, such as those of the hedge-accentor, tree-pipit, brown wren, chaffinch, and greenfinch, no trace of it was found, though the birds sang well; but in others, and especially in those of the starling, sedge-warbler, redstart, blackbird, skylark, and robin, more or less imitation was apparent. About two years later I began to make extensive records of the imitations sung by wild birds, my method being to identify the singer, and then to write down the date and place of singing, the state of the weather, number of times of singing per minute, and the particular sounds which appeared to be reproduced. Some of the sounds imitated were rendered with wonderful fidelity by the better mimics, such as the sedge-warbler, starling, and thrush; other birds, with full-toned voices, as the robin, blackbird, and nightingale, appeared to be able to repeat exactly the musical intervals sung by other birds, but to be unable to render the sounds (especially in the case of harsh sounds) with accurate intonation. Again, some of the singers appeared to whistle their own songs, and then to supplement these with the addition of sounds borrowed from the cries of other birds.

The blackbird and missel-thrush exemplify this method. But the blackcap, whose voice is as full-toned as that of either of these two, adopted an opposite mode, and commenced its songs with soft imitations of other birds, to conclude them with its own full whistles. I attempted to draw no deductions from my first observations, knowing that general statements cannot justly be founded on the behaviour of a few individuals.

During a period of eighteen months I went on with my work of recording, but at considerable personal inconvenience. My profession demanding all the middle hours of week-days, there were comparatively few opportunities for making notes at that time; but on nearly every fine morning, the whole year through, I was early abroad in the woods and fields, armed with telescope and notebook; and when the evenings were long enough, the same course was pursued. Occasionally, my notes were all entered in a ledger, as though to the account of the several birds observed. In the spring of 1889 I began to condense the results of this work, and found myself in the possession of records of the songs of a large number of our commoner birds, heard in Gloucestershire and also in or near Bath, Weston-super-Mare, Clifton, and Bournemouth. I had notes on the songs of some seventy thrushes, nearly sixty robins, many starlings, skylarks, finches, and other singers. A careful comparison of the records revealed that imitative birds had reproduced the cries of other birds especially frequent in their haunts. The skylark had been partial to field-birds; the robin to those of the thicket; the starling to those of the field as well as to those of the town. On further tabulation the notes betrayed indications that summer visitors to this country are imitated more often in spring and summer than in autumn and winter, and this would seem reasonable to any one sufficiently familiar with song-birds to understand that many of them are creatures of intelligence, endowed with accurate memories and considerable powers of observation and mimicry. The various aspects of imitation, and instances of its occurrence, are discussed in the chapter on "The Influence of Imitation," towards the end of this book. In pursuing my studies of the notes of birds, I observed a good deal of bird-life, and soon found that certain cries were employed by birds as call-notes, and others for the purpose of expressing alarm—facts which have long been familiar to ornithologists. I also noticed that birds of a species would generally behave in much the same manner on the same occasions, and that particular cries were sometimes employed to express certain degrees of emotion, some call-notes being evidently of a more urgent meaning than others, and some alarm-cries being similarly more important. There were other curious features in these call-notes and "alarms," which are respectively stated in Chapter II., on "Alarm-notes," and Chapter IV., on "Call-notes."

Between these chapters will be found one treating of the influence of combat. This feature is difficult to discuss with anything like precision: incidents have been adduced tending to show that the pugnacious instinct has some influence on the use of the voice.

I soon found that young birds acquire first the call-cries and alarm-notes of their respective species; that in each species these notes are much less liable to vary than are the songs; and that in different species physically allied, they are more alike than are the songs of those species. Another most interesting feature, and one which I commend to the attention of ornithologists everywhere, was the prominent occurrence of a particular cry in one species; its occurrence in a less marked form in some one or two allied birds, in which another cry might be the most pronounced; and the utterance of this second cry by some other allied birds, which had not the first-mentioned note. I could trace this to some extent in the thrushes, but more especially in the finches. These facts will be appreciated by naturalists, as bearing on the question of a common ancestry of species. For purposes of suggestion I have traced family resemblances of this nature in species physically allied, and in others not thus related. I also read of certain birds which inherit their songs, and I was acquainted with several which inherit their alarm-cries and call-notes. The subject to which these matters relate, the influence of heredity, is discussed at length in chapter vii. There can be no doubt that heredity is as certain in perpetuating the cries of some species as is imitation in determining the whole character of the songs of others; the extent to which these influences prevail in the several orders of birds has yet to be traced. Variation, towards which feature the males of so many species tend, seems to be the parent of imitation; it is considered in a separate chapter.

"The Music of Bird-song" is the concluding theme. It has to some extent been discussed in the chapter on Imitation; but in the Appendix will be found a transcript of many strains which I heard sung by the better singers.

The reader will observe that I have not relied wholly upon my own observations, but have quoted those of numerous well-known authorities, of good report in the scientific world. I read no book at all on the subject until 1889, when I had commenced my comparisons between the various records. Since that time very many works have been perused, but only in a few have any acceptable records been found. Even Mr. Simeon Pease Cheney's work, full of musical notation, hardly touches upon imitation, and has no observations on the influence of heredity. There are poetical writers who describe such incidents as the lark soaring in the sky, pouring out his soul in music for the little brown mate trustfully listening in her nest; but they never remark that the lark utters a chattered song when he fights. There are many ornithologists who name certain birds as imitative; but when their descriptions are examined, the reader can only infer that the mimics were caged specimens. I have nothing to say about caged birds, except in quotation, or in relation to experiments with young birds, or where such arbitrary notes as those of the collared turtle-dove, young pheasant, and young partridge are concerned. In making my investigations, various ideas on the causes of certain features of the exercising of the voice in birds occurred to me. These are set out in chapter vi., on "Noticeable Incidents connected with Bird-song." One of the most marked of these features was the construction of certain songs by the rapid repetition of call-notes.

This subject seems to me sufficiently important to warrant its elucidation in a separate chapter (v.), on "The Simplest Songs."

The scheme of this work is as follows :—A hypothesis on the first occurrence of voice in any animal is stated, and the influence of combat in perpetuating it is then mentioned. The inherited distress-cries of young animals (probably produced long after the occurrence of the voice in adults), and the retention of these cries for the purposes of call-notes, lead to the consideration of the simplest songs, which are mere repetitions of the call-notes, and to those in which variations occur, and which probably are affected by the influence of imitation. The purposes of imitation are then discussed. The scientific value of the notes of birds, as bearing upon the ancestry of species, is considered at length.

By "bird-song" I mean the whole range of the voice in birds. Songs I have defined as vocal utterances, not being alarm-cries or call-notes.[1]

The word "phrase" (as I have employed it in this work) does not mean a strain in music, but a period of song; so that, if a bird sang a few notes at one time, then paused and sang them again, and

[1] *Zoologist*, July 1890.

so on, each repetition would be a "phrase." Thus it is possible that many phrases could be the same song exactly, repeated at intervals, or the song might be varied, but it would be a separate phrase at each time of utterance. It is important that this definition of the phrase should be remembered.

A "strain," on the other hand, means a succession of sounds uttered in a definite order, and which may be repeated, perhaps several times, in one phrase, as commonly occurs in the very long phrases of the skylark.

CHAPTER I

THE ORIGIN OF THE VOICE

THE great diversity of the voice in animals, and the extent to which it is inflected by individuals, suggest that it is particularly liable to be influenced by the vicissitudes of the struggle for existence, and to be necessarily modulated in accordance with the requirements of each animal or species by which it is employed.

This modulation is very apparent. We hear from every side the notes of various animals, as cries for succour, songs apparently of love, and notes equally suggestive of triumph or of fear— all of them exactly suited to the purposes for which they are uttered. But it does not follow that these sounds have always been employed for the same purposes, nor that they have always been uttered in the present modes; for we observe among birds, at least, prevalent changes not only of musical pitch, but of other vocal characters, in the young as well as in the adults. Nor must we hastily assume that the voice itself has always been coeval with even the higher forms of animal life upon the earth. On the contrary, there are grounds for believing that in the epoch when the highest forms were the amphibians and reptiles, the voice, properly so-called, had not yet been heard in the world.

Charles Darwin was of the opinion that the first vocal sounds were involuntarily produced. In his work, *The Expression of the Emotions*, pp. 83 and 84, is the following observation: "When the sensorium is strongly excited, the muscles of the body are generally thrown into violent action, and, as a consequence, loud sounds are uttered, however silent the animal may generally be, and although the sound may be of no use. Hares and rabbits, for instance, never, I believe, use their vocal organs except in the extremity of suffering; as when a hare is killed by a sportsman, or when a young rabbit is caught by a stoat. Cattle and horses suffer great pain in silence, but when this is excessive, and especially when associated with terror, they utter fearful sounds. Involuntary and purposeless contractions of the muscles of the chest and glottis, excited in the above manner, may have first given rise to the emission of vocal sounds."

The most violent contractions of the muscles of the chest and glottis would certainly have occurred during combat, and it is therefore fair to assume that the voice was first produced during the fights of animals, although possibly its earliest tones were little more than coughs or grunts, or a hoarse murmur due to panting. Darwin observed (*ibid.*), " The principle also of association, which is so widely extended in its power, has likewise played its part. Hence it follows that the voice, from having been habitually employed as a serviceable aid under certain conditions, inducing pleasure, pain, rage, etc., has commonly been used whenever the same sensations or emotions are excited under quite different conditions or in a lesser degree."

We may be certain that the perception of a threatened attack is only a mental anticipation of the combat which experience, or inherited instinct, suggests as a probable consequence of the approach of an enemy. It is easy to imagine that, among animals which survived by the agency of their speed, the excitement due to the approach of an enemy would cause an involuntarily increased rapidity in the movements of the lungs; and if the animals were of large size, a kind of snorting, like that of a horse, might thereby be produced; or perhaps the initial movement of certain limbs, like the twitching of the wings of a terrified falcon, would be occasioned. The development of clear cries from snorts or grunts must be regarded as a process of very slow progress. But we must remember that there are several widely distinct races of animals whose only attempt at vocal utterance is the employment of a yet simpler mode of expression—a mere puff or hiss caused by the toneless expulsion of air from the lungs; and be it observed that in these instances, as well as in those of its occurrence in animals possessing voices, the sound is invariably due to anger or to fear. The common tortoise, so often kept as a pet in English gardens, when suddenly alarmed withdraws its head and limbs rapidly within its shell, and at the same time a kind of short hiss is heard, especially if the animal has been frightened by being grasped in the hand. In serpents, a hiss is the common expression of anger. Among birds, the same feeling is sometimes expressed in a similar tone, but it is rarely employed except by birds sitting on their nests. In this position, the female nuthatch, the wryneck, and the blue titmouse, utter a hiss not unlike that of a snake—a circumstance well known to egg-collectors. The brooding domestic duck hisses at the hand which attempts to remove her eggs; and, under the same conditions, the common domestic hen makes a violent puffing, less harsh in tone, but otherwise greatly similar to the hissing of the duck.[1] The great owl (*Bubo maximus*), when angry, utters a sharp, loud hiss, not unlike the sound produced by the common brown owl of England when irritated. I have only observed one great owl, a bird two years old, in very comfortable quarters near Stroud. The young hoopoes, when disturbed on the nest, "crowd forward and utter a hissing noise" (Harting, *Summer Migrants*, p. 255). The gander, too, hisses as a means of menace to any one threatening his goslings (Jesse, *Gleanings*, p. 48). The mute swan hisses loudly at intruders when his mate is on the nest. The nestling young of the common pigeon and those of the turtle-dove (*C. turtur*) make a blowing or puffing sound as a menace. The common rosy cockatoo employs a short, sharp puff for the expression of anger; and I heard a precisely similar sound uttered by two fresh-

---

[1] A few years ago, in May, I saw at the Zoological Gardens in Regent's Park, London, a ruddy shelduck chasing a gull around his pond. The shelduck was hissing violently.

caught and hooded peregrine falcons in the possession of Major C. H. Fisher.

The angry hissing or puffing of birds has not been recorded by observers; but in the above list are birds of widely distinct genera, induced by inherited tendencies to utter sounds which, in a general sense, may be considered as restricted to the reptile kingdom. This seems the more important when we remember how closely the bird is physically related to the reptile.

Whatever may have been the character of the first vocal efforts, it is interesting to observe that at the present time there are creatures of common occurrence in which the voice is of an entirely accidental origin. The appearance of these animals does not suggest a tendency towards song, for they are none other than the three commoner forms of newts found in Britain, namely, the triton, the smooth newt, and the palmate newt — creatures allied to the salamanders, frogs, and other amphibians. They commence life as tadpoles, and afterwards come ashore to spend a period of some four years in developing towards maturity. At the end of this period they resume an aquatic existence for the purpose of continuing their species. They breathe in the manner of frogs, forcing air into their lungs by an action akin to swallowing, and retaining it there by means of the closing of the glottis. After spending some months in the water, they emerge once more, to stay on land for the remainder of the year. But during all this time they never utter any sound which can be considered as vocal, though, in the act of breathing, their jaws sometimes close with a sort of snap. Persons who have kept newts in aquaria are generally aware that, when a newt is held by the tail it wriggles violently, and often, when so struggling, utters a brief but distinct croaking sound. The noise is, no doubt, caused by the body of the creature striking against the fingers, by which action the lungs are compressed, and consequently the contained air is forced through the tightened glottis, and thus produces a croak. Perhaps the gigantic ancestors of our newts did not struggle quite so actively, but any sounds, proportionately louder, thus produced in their contests, may have been of service by occasioning the release of amphibians from the grip of predacious enemies to whom such outbursts were unfamiliar. When a newt is seized towards the hind quarters by a snake the croak is heard; but I cannot say that I have ever known it to be of any avail against such an attack, and I have witnessed the incident some hundreds of times. Young frogs (*R. temporaria*) sometimes scream when seized by a snake; and in this case an inherited cry is produced in its full tone by an animal which has never before made use of any vocal power it may possess.

Among the earlier forms of life the voice was probably of accidental rather than voluntary origin. In the amphibians it may have been originated in the same manner as it is now first used by the newts, and have survived in consequence of an advantage it conferred in battle. Among mammals and birds it may have been inherited from an earlier source; or in the former it may have been elaborated from snorts and grunts, and in the latter the violent actions and rapid breathing incidental to combats in the air may have initiated a power which was afterwards adopted as a means of expressing a threat. Birds of several genera make an audible snapping noise with their bills when they fight; but the greenfinch, chaffinch, and house-sparrow, under the same conditions, sometimes produce a hoarse kind of note which appears to be due partly to the very rapid striking together of the mandibles, and partly to some guttural sound. This reference to the house-sparrow must be read in connection with subsequent remarks upon the so-called noisy combats of this bird, which are not fights at all, but affairs of love. When sparrows really fight they make but little noise, and then the sound just described can be distinctly heard by an observer who is fairly close to the combatants.

It is to be regretted that the sounds made by fighting birds have been, like those uttered by birds disturbed on the nest, so little investigated or recorded by ornithologists. If Darwin's theory, that the voice was first occasioned by the involuntary contraction of muscles, consequent upon excitement of the sensorium, is sound, it is equally true that the greatest excitement of the sensorium is occasioned by the attack or the mere presence of an enemy (actual or supposed); for several animals, such as cats, rabbits, and rats, will remain quiet when in a trap, although enduring great physical suffering, but, if approached by a person, they begin to make an outcry. The same behaviour is noticeable in many birds; and in the reptile kingdom we often observe the only attempts at vocal utterance to be intended as expressions of intense anger or fear.

We may consider the voice to have been evolved from a toneless puffing, indicative of anger, or from snorts or grunts accidentally caused; and if, on the other hand, we accept Darwin's view, we can, not the less reasonably on that account, hold to the opinion that the voice was first occasioned by the agency of combat.

CHAPTER II

ALARM-NOTES

IF any particular kind of vocal utterance had been of aid in combat, it would, in a comparatively short period, have become general in the species by which it was employed.

A combat-cry might well be uttered before the commencement of a fight, and it would then be virtually a defiance, and, in the neighbourhood of assembled animals, it might answer the purposes of an alarm-note. We may be sure that such a cry would have been acted upon, since, among warm-blooded animals, alarm-cries are of almost universal comprehension. Birds in a thicket become suspicious when some distant individual of even another genus utters an alarm. I remember a marked instance of this habit, in which several birds of distinct genera—tits, nuthatches, goldcrests, and tree-creepers—gathered around a coaltit which, alarmed at the sudden appearance of a little rough terrier, was uttering loud simple screams; the others hurried to the spot and joined in a chorus of alarms.

When a bird is greatly excited by fear, either for its own safety or on account of its young, and generally when other birds are near, it usually utters its common alarm-cry many times in succession. In some species certain alarm-cries consist of a rapid repetition of a single note (the alarms of the mistle-thrush and magpie are examples), and in other birds they are single cries repeated at intervals, as in the nightingale, chiffchaff, chaffinch, and crow. But sometimes a bird with a single alarm-note will repeat this cry so many times in rapid succession that a kind of exclamation is produced of the same character as the set or arbitrary alarms of the mistle-thrush and magpie.

Thus, I have heard a nightjar utter its sharp, clicking alarm at first slowly, then more and more rapidly, until the notes became blended into one rattling cry. I have twice heard a magpie similarly utter short *cahs*, and rapidly increase their number, slightly changing their tone, until the common *shŭshŭshŭshŭ* alarm-cry was rendered in its accustomed style. On several occasions I have also heard a male house-sparrow begin his continuous shaking alarm by uttering his common call or note of warning and occasional song-note *tell;* and here also the same increased rapidity of utterance seemed to have the same result.[1] On many occasions I have heard a male blackcap, alarmed for the safety of his young, repeat the ordinary alarm *tack* (well named by Yarrell) so rapidly that a continuous shaking cry, like the shaking alarms of the house-sparrow and great tit, was produced. I have twice heard long-tailed tits reiterate their rattling alarm in the manner in which the call-note *tuck* of the brown wren (a cry most often heard in autumn) is very frequently prolonged, so that the utterance was changed like that of the above-mentioned blackcaps. On one occasion the tits were frightened by a pair of sparrow-hawks flying low above them, and on the other they were alarmed by the sudden appearance of a strange dog. It is a matter of easy observation that the rattling alarm of the blackbird is constructed of repetitions of one cry, which is sometimes uttered slowly at intervals of several seconds; but this is only when the bird is suspicious of some concealed danger, and as the peril increases the notes become more frequent, more acute, and considerably higher in pitch. This incident affords a

[1] The female house-sparrow has both of these cries.

common illustration of the general rule that when a vocal sound is employed for the expression of an emotion, and the emotion becomes intensified, the note is uttered more frequently, and at a higher pitch. Of course the blackbird is often alarmed suddenly, and then he utters his full alarm at once; but when his young are threatened, as it seems to him, by the presence of another animal, the construction of the full alarm-cry is sometimes beautifully demonstrated. If the rapid alarms here mentioned have been originally produced by mere repetition, is it not probable that alarms which invariably consist of a note repeated many times, such as those of the mistle-thrush and the magpie, were evolved in a similar manner?

Danger-cries were probably often uttered near the nests of many species, as they now are thus employed by the lapwing, blackbird, house-sparrow, chaffinch, willow-warbler, and many others of the same genera; and they would also have been uttered near the fledged young when these were threatened; hence the young would have been familiar with these cries, and in due course would have reproduced and perpetuated them. In this way each race would probably have developed its own danger-cry; for we may be sure that there was variation, if only in degrees of vehemence, and consequently in length or rapidity of repetition, in these cries. The blackbird affords an example of such an occurrence. Its alarm is very like the alarms of its allies the ring-ouzel, mistle-thrush, and fieldfare, but is uttered more frequently, more rapidly, and with greater variety of pitch than any alarms of those species; and sometimes, especially towards evening in February and March, it appears to be employed for the mere purpose of making a noise and attracting attention. The influence of heredity in perpetuating alarm-cries will be discussed in later pages.

The relation of particular alarm-cries of birds to the presence of certain enemies is a deeply interesting subject, which has not received much attention from ornithologists. Many birds, when approached by enemies, express vocally, as well as in other ways, various degrees of fear, which indicate a considerable amount of discrimination. The expression of fear may have been inherited, as it is in the domestic cock, who yet retains the prominent habit of uttering a yell of alarm when any large bird unexpectedly flies over him, a habit exhibited by his remote ancestor (*Gallus bankiva*) in India.[1] Mr.

[1] It is worthy of remark that often, when first the young cockerel attempts to crow, he utters a simple yell, closely similar to the alarm-yell of the adult bird.

W. H. Hudson has recorded the discrimination of the birds in the La Plata region, where the various hawks and falcons "have just as much respect paid to them as their strength and daring entitle them to, and no more" (*Naturalist in La Plata*, p. 83).

Various incidents have been recorded which seem to suggest the relation of cries to the presence of particular enemies. In *Traité de la Fauconnerie* (by Prof. Schlegel), p. 44, is the following: "The signs of alarm which the shrike (*L. excubitor*) gives vary infinitely, not only according to the species of bird of prey which appears, but also according to the mode by which it approaches—whether slow or quickly, gliding over the ground, or soaring aloft, and so on."

The writer of this observation refers to a caged shrike kept to announce the arrival of a hawk in the vicinity of the trapper. My own experience has been sufficient to enable me to generally gather from the alarm-cries of birds a fairly correct impression of the nature of the enemies they announce. I know well that when house-sparrows utter loudly the most vehement alarm of the species—a sound resembling that of *tourr*—the birds are greatly afraid of some other bird, and that probably a hawk is in sight. But I have known them utter the same sound when a jackdaw attacked their young. I have often discovered the presence of a hawk from hearing the starlings utter sounds which I will term their " hawk-alarm," namely, cries sounding like *cack-cack*. The starling has other notes also employed to express alarm ; but this cry is certainly of general and almost exclusive use when a hawk suddenly appears. But I cannot say that it announces any particular species of hawk, for I have heard it employed by starlings which were approached by a tame kestrel, and by others at which a tame peregrine falcon swooped ; and the wild sparrow-hawk is always announced by this exclamation.

Blackbirds utter a loud metallic chirp when their nests or young are approached by a cat, or when they find an owl. A friend informed me that he observed in his garden a number of blackbirds uttering this cry very noisily near a cucumber-frame, in which a female blackbird had accidentally become enclosed. The captive was liberated, and the outcry immediately ceased. I trained a little Dandie terrier to recognise the cries which blackbirds in the garden uttered when a cat was near their nest. This was a frequent occurrence ; and as soon as the birds began to rate their enemy, the dog would run into the bushes and compel it to beat a hasty retreat. The birds took no notice of the dog.

The swallow doubles its common call-note *clit* into *clit-it* when a hawk approaches. But I have heard it utter precisely the same cry when its young were attacked by a crow. I have noticed this behaviour in swallows at Vancouver, B.C., as well as in those of England. The common domestic pigeon announces the arrival of a hawk or a hawk-like bird by a peculiar puff or grunt, something like the sound of the word *oof;* but it employs the same sound when disturbed by a man at night. The common cock, when alarmed by a bird flying over him, utters a loud yell, as has been noticed; but, if approached by a bird or other animal upon the ground, he almost invariably utters a cackle of surprise.

It appears, then, that certain birds utter particular cries to express great fear, whether the fear be caused by the actual presence of a much-dreaded enemy, or by an erroneous impression received from the similar movements of some other creature. But we are not at present in a position to assert that any particular enemy is announced by any particular cry. There is, however, no occasion to doubt that parent birds, partly by means of cries as well as by deportment, educate their young in a knowledge of their rapacious enemies, although in certain species this knowledge is to some extent inherited, as it is in the common fowl. In relation to this interesting matter the following remarks by Mr. Hudson are in point :—" Another proof that the nestling has absolutely no instinctive knowledge of particular enemies, but is taught to fear them by the parents, is to be found in the striking contrast between the habits of parasitical and genuine young in the nest and after they have left it, while still unable to find their own food. I have had no opportunities of observing the habits of the young cuckoo in England with regard to this point, and do not know whether other observers have paid any attention to the matter or not, but I am very familiar with the manners of the parasitical starling or cow - bird of South America. The warning cries of the foster-parent have no effect on the young cow-bird at any time. Until they are able to fly, they will readily devour worms from the hand of a man, even when the old birds are hovering close by and screaming their danger-notes, and while their own young, if the parasite has allowed any to survive in the nest, are crouching down in the greatest fear. After the cow-bird has left the nest it is still stupidly tame ; and more than once I have seen one carried off from its elevated perch by the milvago-hawk, when, if it had understood the warning cry of the foster-parent, it would have dropped down into the bush or grass and escaped ; but, as soon as the young cow-birds are able to shift for themselves and begin to associate with their own kind, their habits change, and they become suspicious and wild like other birds" (*op. cit.* p. 89).

It would seem that the alarm-cries of birds were at an early period developed so as to induce silence in the objects to which they were addressed. Mr. Hudson, who has carefully considered this subject, says: "When very young, and before their education has well begun, if quietly approached or touched, they open their bills and take food as readily from a man as from a parent bird; but if, while being thus fed, the parent returns and emits the warning-note, they instantly cease their hunger-cries, close their gaping mouths, and crouch down frightened in the nest. This fear caused by the parent bird's warning-note begins to manifest itself even before the young are hatched; and my observations on this point refer to several species in three widely-separated orders. When the little prisoner is hammering at its shell and uttering its feeble peep, as if begging to be let out, if the warning-note is uttered, even at a considerable distance, the strokes and complaining instantly cease, and the chick will then remain quiescent for a long time, or until the parent, by a changed note, conveys to it an intimation that the danger is over" (*Naturalist in La Plata*, p. 89).

I have during many years observed with some amusement the warnings by which, at daybreak, old house-sparrows silence their impatient young. The sparrows have been accustomed to sleep, with their broods, among creepers affording excellent shelter immediately beneath the open window of my bedroom. At morning twilight the young birds would begin to cheep, and the old ones had some difficulty in keeping them quiet. This they effected by uttering various (and certainly, as they seemed to me, rather suggestive) inflections of the favourite note of warning and occasional song-note, *tell*. At intervals of a minute or two the young would recommence their importunate cries, and an old bird would at once impose silence by a sharp note, followed by various *sotto voce* admonitions.

CHAPTER III

THE INFLUENCE OF COMBAT

THE probability that combats between animals first occasioned the utterance of vocal sounds has been suggested in Chapter I. of this book. The purport of the present chapter is to discuss the further influence of combat in developing and modifying the voice. Just as an anticipation of danger may produce an alarm-cry, so the irritating presence of a rival, or an intruder upon a favourite resort, might have induced the use of a cry as a defiance.

Darwin observed : " It is certain that there is an intense degree of rivalry between the males (of birds) in their singing. Bird-fanciers match their birds to see which will sing longest ; and I was told by Mr. Yarrell that a first-rate bird will sometimes sing till he drops down almost dead, or, according to Bechstein, quite dead, from rupturing a vessel in the lungs (*Naturgeschichte der Stubenvogel*, 1840, p. 252).

That the habit of singing is sometimes quite independent of love is clear, for a sterile hybrid canary-bird has been described as singing whilst viewing itself in a mirror, and as then dashing at its own image; it likewise attacked with fury a female canary when put into the same cage. The jealousy excited by the act of singing is constantly taken advantage of by bird-catchers: a male, in good song, is hidden and protected, whilst a stuffed bird surrounded by limed twigs is exposed to view. In this manner, as Mr. Weir informs me, a man has in the course of a single day caught fifty, and in one instance seventy, male chaffinches. That male birds should sing from emulation as well as for charming the female is not at all incompatible; and it might have been expected that these two habits would have concurred, like those of display and pugnacity. Some authors, however, argue that the song of the male cannot serve to charm the female, because the females of some few species, such as the canary, robin, lark, and bullfinch, especially in a state of widowhood, as Bechstein remarks, pour forth melodious strains."

It may be presumed that these latter individuals were caged birds, having abundant leisure and plenty of food, incidents the effects of which I shall venture to discuss in subsequent pages. Darwin continues:

" In some of these cases the habit of singing may be in part attributed to the female having been highly fed and confined, for this seems to disturb all the usual functions connected with the reproduction of the species. . . . It has been also argued that the song of the male cannot serve as a charm, because the males of certain species, for instance of the robin, sing during autumn. But nothing is more common than for animals to take pleasure in practising whatever instinct they follow at other times for some real good" (*Descent of Man*, p. 369). Singing affords almost the only means of occupying the ample leisure of a captive bird.

Amongst polygamous species the selection of mates is largely dependent upon victory in combat; and this may account not only for the defiant nature of the cries of male Gallinaceous birds, but also for their lack of varied song. The males capture rather than woo their mates: hence sweetness of voice is of far less advantage than strength and fury in their courtship.

It is clear that a widely-prevalent snapping noise made with the bill has been developed among birds as a menace; and it corresponds to the display of teeth by an angry dog, and the waving of the horns and lowering of the head of an enraged bull. I have not detected any similar snapping of the bill during song, unless, indeed, a certain note of the greenfinch can be considered to be, as its tone indicates, modified by a rattling of the mandibles during its utterance. The full song of the male greenfinch may be rendered as follows: *Did it it ititit, tell tell tell, zshweeoo,* the *zsh* in the last word being pronounced like the letter "s" in "treasure" and "pleasure." The cry *did it it it* is the common call of both sexes at maturity. About February the male generally adds to this the cry *tell tell* . . .; and in March is added the *zshweeoo*, which is never uttered before *did it* or *tell* in the phrase, but often follows them, and is sometimes uttered alone. This note is never uttered when the male is "playing" to the female, nor when he is feeding her off the nest, nor yet during flight; but it seems to be accompanied by a rattle of the bill closely like that which is produced during combat by this species, and by the female house-sparrow when she threatens the male who is courting her. For these reasons I consider this note to be a defiance song-note. A very similar, but shorter and much coarser, note is uttered by the goldfinch when fighting; and the lesser redpoll and (I believe) the siskin furnish similar examples.

There is, however, much better evidence that defiance influences the vocal utterances of certain birds.

The cock does not crow when calling a hen towards him, but invariably crows when he has defeated an adversary, and generally before combat. The males of the American pinnated grouse (*Tetrao cupido*) utter a "cackling, screaming, and discordant cry when fighting at their scratching places, where they meet at break of day" (Wilson, *Amer. Ornith.*, vol. ii. p. 399). The kingbird (*Tyrannus Carolinensis*) twitters when attacking the eagle (Wilson, *op. cit.* vol. i. p. 218). Of the snow-bunting, Saxby stated, "They chirp as they fly, and the sudden jarring sound is heard which is uttered by the bird when suddenly directing its course towards a neighbour. That the note in question is sometimes one of anger I have repeatedly observed, when two of the birds have been quarrelling over their food" (*Birds of Shetland*, p. 91). I have often observed that in early spring yellow buntings and larks utter a kind of chattering song when they contest among themselves the possession of favourite spots. These birds are then appropriating situations for nesting, and are seeking mates. Twice I have heard chaffinches cry *pink pink*, and that cry only during combat. The house-sparrow is generally silent when fighting. The so-called noisy combats of this species are simply assemblies of males around a female, to whom they are showing-off and chirping, and from whom they often receive violent pecks and pinches which sometimes make some of them scream. When males fight, one may hurt the other and cause him to cry out. This one then exclaims in precisely the same tone as that employed when a sparrow is carried away by a hawk; but until so injured by an opponent he is silent. The females sometimes severely pinch males which play to them too demonstratively: I have twice seen a male hanging suspended for certainly two seconds by the bill of a female.

The crested lark (*Alauda cristatus*), like the skylark, "has the peculiarity that when it fights it continues to sing" (Bechstein's *Nat. Hist. of Cage Birds*, p. 179). I have heard the full song of the tree-pipit repeated by this bird when fighting furiously. During spring and summer male thrushes sing when fighting, and during the whole year, excepting July and August, male robins have the same habit; in both species the song becomes raised in pitch, often to an obscure squeak, before the combat commences. Jesse observed this habit in the robin (*Gleanings*, 2nd series, 1834, p. 252), and he recorded that a blackbird, after beating a cat away from its young, celebrated the victory with a song (*tom. cit.* p. 248). The hedge-accentor twitters when fighting. I have often observed chiffchaffs and willow-wrens, and occasionally goldcrests, singing while fighting. In spring brown wrens challenge each other in song. In *The Zoologist* for 1869 (p. 1645) is an account of brown wrens which were fighting, and after being separated three times, "sounded notes of battle and began to fight again." I have twice similarly separated fighting wrens, and on each occasion the victor immediately began to sing, and in his song omitted to utter the coarse call-note, which is generally included.

Darwin says, "Rival males try to excel and challenge each other by their voices, and this leads to daily contests. . . . Thus the use of the voice will have become associated with the emotion of anger" (*Expression of the Emotions*, p. 85). It is possible that rivalry may sometimes induce definite defiance, as in autumn it appears to do in robins, which, after singing near each other, approach and fight. Some observers seem to have been of this opinion, for Darwin states, "Many naturalists believe that the singing of birds is almost exclusively the effect of rivalry and emulation, and not for the sake of charming their mates" (*Descent of Man*, p. 369). "Males of song-birds and of many others do not in general search for the female; but, on the contrary, their business in the spring is to perch on some conspicuous spot, breathing out their full and amorous notes, which by instinct the female knows, and repairs to the spot, to choose her mate (Montague, *Orn. Dict.*, ed. 1833, p. 475). Mr. Jenner H. Weir informs me that this is certainly the case with the nightingale" (*Descent of Man*, p. 368). This seems to be borne out by a later observation. "Bechstein, who kept birds during his whole life, asserts 'that the female canary always chooses the best singer, and that in a state of nature the female finch selects that male out of a hundred whose notes please her most' (*Naturgeschichte der Stubenvogel*, 1840)." It would seem that Bechstein, in making this assertion, was assuming a familiarity with the feelings of wild birds very difficult to acquire; for how can an observer be sure that such a preference is not due to colour or movement, instead of to song? But probably before there was a call-song there were heard varieties of call-notes; and this subject will be discussed in the next chapter.

CHAPTER IV

THE CALL-NOTE

IT may fairly be said that any bird-cry which induces a bird to approach another is a "call-note," whether it be a danger-signal, a combat-cry, or an alarm. But naturalists have distinguished in nearly every kind, and certainly in all British kinds of extended song, specific notes which they have observed to be used only for the purpose of attracting birds; and these notes they have termed "call-notes." It is true that different writers have sometimes mistaken the nature of such cries, and have not only given conflicting descriptions of call-notes, but have also considered some battle-cries to be of this character. The two following instances are examples:—The call-note of the marsh titmouse (*Parus palustris*) has been variously described, but no description is better than that of the late Mr. Sterland. In his *Birds of Sherwood Forest*, p. 86, he says of this species: "Their cry resembles the words *chica, chica, chica*, repeated four or five times in succession, and ending with a shorter syllable, *chike*." In Gloucestershire the principal spring call-note is a rapidly downward-slurred whistle, sometimes followed by the quick repetition of a harsh cry, the whole sounding something like this (the actual pitch is wholly immaterial):

Cheeū sa sa sa Cheeū Cheeū Cheeū sa sa sa sa.

This seems to me the only characteristic call-note of the marsh titmouse. The well-known cry *fink* or *pink* of the chaffinch has often been termed the call-note of the bird, and even Yarrell has made this mistake. My experience convinces me that the cry is not a call-note. The chaffinch has several notes which are of this character: one is a soft cry almost exactly like the ordinary *chissick* of the house-sparrow, and is uttered by both sexes at the breeding season; this also is the call-note by which the young attract their parents. Another cry which I have often heard uttered by male chaffinches before the breeding season is a loud short whistle very rapidly slurred upwards, in the interval of about a fifth or sixth. It may be pronounced *twit*, and written thus :

In autumn yet another cry is employed, a very short and soft chirp or whistle, descending in the interval of a whole tone. The cry *pink*, which is quite distinct from the other notes, is a battle-cry and alarm. I have twice in winter heard it used by fighting chaffinches, and on several other occasions it has been accompanied by a menace. On one occasion (in March) I saw two male chaffinches flying around a female. The leading male, evidently trying to avoid the attack of his pursuer, uttered the soft call-note (like *chissick*), seemingly as a call to the female. The pursuing bird uttered *pink pink* very loudly. This incident lasted for about one minute. I have also heard the latter cry uttered by caged chaffinches when much alarmed. Mr. Herbert C. Playne informs me that this cry is frequently employed when chaffinches are mobbing an owl ; I have heard it employed by them towards my tame kestrel ; and Rennie's *Domestic Habits of Birds* (p. 248) recorded that some chaffinches cried in the same tone at a pine-marten. This note is also uttered in autumn and winter, more especially in the early morning hours; but, notwithstanding this fact, the note appears to be less often used than any other where a call-note might be employed.

The first call-notes of birds were probably mere adaptations of alarm-cries, the use of which was induced by the influence of mutual aid among associated individuals, by which the danger-cries would have been frequently employed. But the necessities of the young, having relation not so much to danger as to hunger, must have early brought about the use of cries for food. These simple cries, from having been effective in result, would also have been used by the young after leaving the nest, and might then have indicated a desire not only for food, but also for the companionship of the parents. Under these circumstances such cries would have become call-notes.

It is probable that the first cries of the young were merely inherited modes of expressing distress, and were not influenced by mimicry; for we now find that the young of many species, of limited vocal range, utter cries which are wholly unlike those of their parents. This occurs in several rasorial birds, and others.

The young of birds which rear large broods would

make frequent use of their call-notes; for, as the assembly moved about in the search for food, some of them would often be left behind, when they would call to the others, and be answered, at least by the parents. Most birds which rear large broods are noisy when with their young; thus the titmice, goldcrest, brown wren, starling, finches, and others make abundant use of their calls towards their broods; and those of them which continue the association throughout the summer likewise retain this vocal habit. In fact, among birds, as in man, language is very closely related to the social instinct, notwithstanding that its development, among certain gregarious species, seems to be so much retarded by fear.

Various emotions have predominated in different species, to the exclusion of sociability or of garrulity. The lack of variety in the voices of the larger birds may be accounted for on the supposition that fear, unless acute, induces silence, and that these larger species, ever seeking concealment from enemies, could not without incurring grave danger make frequent and loud noises. Of course, in the breeding season, the emotion of fear is frequently overcome by that of love, and then birds often expose themselves to danger while singing. The most marked instance of this which has come under my observation occurred when a nightingale allowed me to approach within three feet of him, without interrupting his song—he being apparently as ignorant of my presence as of the keen delight I derived from his music. Hunger, again, may induce incessant labour, like that of the woodpeckers; or watchfulness, with occasional repletion and consequent lethargy, as in the Colymbidæ, Raptores, and others,—all of which conditions militate against the development of the voice, for they are not consistent with that abundant leisure and contemporaneous vivacity which are necessary to song, either in birds or in man.

The persistence of certain characteristic cries, including call-notes, among species with extended vocal range, may be accounted for not only by the filial mimicry of the nestlings,—and this in many species is very powerful,—but also by the importance of the recognition of the notes by the parents, which are generally directed to their fledged young by the cries of the latter; and we may also suppose that sexual selection has likewise tended to the perpetuation of certain cries, for a female bird would very probably prefer a male whose call-note closely resembled the tone which she, when young, had so often employed to obtain food and protection.

The frequent similarity of the danger-cries and the call-notes among birds physically allied, but whose songs are quite distinct, will be discussed in the chapter treating of the influence of heredity. It may be here mentioned as affording strong evidence of the original development of the call-note from the danger-cry.

Several influences would have determined the particular characteristics of the cries uttered by different races of animals ; among these, the capacity of the throat, due to a habitual mode of feeding, also an ability to repeat a cry many times with one breath doubtless had some effect ; and in songbirds the tones might have been gradually affected by any recurrent prevalent sounds falling on the receptive senses of successive generations of the young. The perpetuation of characteristic notes of birds may be easily accounted for. The recognition of them by parent birds was of vital importance to the young, whose cries for food would have been effectual in proportion to their vehemence and the certainty of their being understood. But if males uttered songs (by which I mean vocal sounds distinct from alarms and call-notes), they might on this account be preferred, and in this event the call-note would be less frequently employed in song by the most extravagant singers. This is obviously the fact. The blackbird, thrush, robin, and blackcap never utter a call-note during song, but they seem to trust only to their more original music. The nightingale often commences its phrases by uttering one of its call-notes. The song of the skylark is often, in early spring, concluded in long notes which resemble the call-note of the young skylark prolonged. It is obvious that the phrases of inferior singers, such as the greenfinch, creeper, yellow bunting, bullfinch, pied wagtail, hedge-accentor, golden-crested wren, nuthatch, and titmice consist more or less of repetitions of the call-notes of their respective species.

CHAPTER V

THE SIMPLEST SONGS

AT the commencement of the breeding season birds of opposite sex call to each other; and this vocal exercise is especially performed by the males. I know of no species (except, perhaps, the mute swan) in which this behaviour does not occur. As the increase of fear induces amongst birds in company a more frequent repetition of an alarm-cry, so love stimulates the male to a more rapid repetition of his call-notes. In some species there is apparently a diversity of call-notes; but it will be observed that generally a particular cry is employed by both sexes. At the commencement of this season, then, the kestrel repeats his cry to his mate; the rook calls loudly and incessantly from the selected site for the nest; the tree-creeper repeats with unusual rapidity his long-drawn and plaintive note; the wryneck cries continually in rapid successive shrieks; the ringed plover then doubles his note (Harting, *Birds of Middlesex*, p. 148); the lesser spotted woodpecker repeats his only note, *tic, tic,* or *krick*, continuously (Yarrell, vol. ii. p. 479); the pine-grosbeak also repeats a call-note "with variations, sufficiently often to attain the dignity of a song" (Yarrell, vol. ii. p. 183); the greenfinch prolongs the repetition of his common call-note, *did it it ititit*, and performs extravagant feats in flight; the bullfinch repeats his common call-note; the hedge-sparrow repeats his call-squeak many times in succession; and even the *chissick* of the pied wagtail is greatly in evidence. Then another phase ensues. The creeper repeats his cry many times without an interval. The greenfinch adds to his common call another call-note, *tell, tell,* or *yell, yell* (as it is better rendered); and he utters these notes in a regular succession, thus producing a phrase which is sung both on the wing and from a perch. It may be suggested thus: *Did it it itititit; yell, yell, yell; did ititit; yell, yell.* The bullfinch varies the repetition of his notes, and thus commences a warble. The golden-crested wren constructs a phrase by uttering his call-squeak twice in double time, afterwards four times in succession, and in the latter stage the pace is accelerated towards the close. The hedge-sparrow repeats his squeak, then two or three very rapid notes at a lower pitch, then his squeak again; and by repeating this performance more than once, he produces a song which embodies several repetitions of his common call-squeak. These repeated squeaks are uttered at the same pitch, and often the phrase is concluded with one of them. A similar method of singing is constantly exhibited by the brown wren, and sometimes even by the robin. The pied wagtail at first departs but little from a mere repetition of his call; but later in the spring he manages to construct a sort of jumbled song, in which, however, his habitual notes still predominate. Thus also the cushat gradually elaborates his song or coo; the stock-dove never exceeds the bare repetition of a short, jerky *coo;* and the domestic cockerel repeats a succession of alarm-yells when first he attains the power to essay a crow. The yellow bunting repeats a note which he often utters—it may be suggested by either *jip* or *jink* —and by means of such repetitions he produces the first part of that song which country folk have likened to the words, *A little bit of bread and no cheese.* The willow-warbler repeats his alarm call-note—a whistle sounding like the word *tewy*, and slurred upwards in this manner—

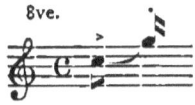

but rendered at a higher pitch. With progressive variations he sometimes constructs from this cry his beautiful song, which can hardly be reproduced in notation, but may be suggested thus:

The above note *tewy* is a call-note of the nightingale to its young, and is uttered by this bird at the commencement of about one-fourth of its phrases. I have often heard it repeated longer, and gradually more drawn-out, until the well-known long notes of the nightingale were produced in their typical character.

Then, also, the nuthatch repeats his full-toned cry *tewit*, which at all seasons we hear about the elms, so quickly that he produces a full, prolonged, bubbling cry, in character resembling the rolling "water-bubble" notes produced by a quill whistle blown in water. The skylark, whose first efforts towards song may be heard in autumn and winter, follows the same method of repeating the call-note for a song—a method which he soon abandons. In early spring the phrases of the starling are concluded with a harsh, squealing cry, which frequent observation leads me to designate its note of passion. This cry is sometimes addressed as the most vehement call to the female, and is also sometimes uttered when she is not near. It is rarely heard in July, August, or September. I have heard the tawny owl repeat his hoot so frequently as to produce a sound like the bubbling note of the nuthatch. Dr. A. G. Butler, F.L.S., informs me that his zebra finches repeat the call-note four or five times as a song.

In all the above-mentioned British species, and in some of their allies, which represent many avian races, the males court the females partly by the repetition of notes which we observe to be employed in other circumstances as call-notes; and in some species these notes are repeated so rapidly that a phrase is constructed. But some species have never advanced beyond the mere repetition of their call-notes. Of such are the mallard, stock-dove, kestrel, blue tit, and others of limited vocal range. Mr. Hudson, in his book on La Plata (p. 258), has some suggestive observations on the inherited simple song of the tinamu. *Rhynchotus rufescens*, he says, has a song or call " heard oftenest towards evening, which is composed of five modulated notes, flutelike in character, very expressive, and uttered by many individuals answering each other as they sit apart concealed in the grass." This appears to be an invariable strain, frequently employed, and transmitted to the young by inheritance. In some birds the call-note is extended, as it is by the nuthatch, which prolongs its full-toned short cry *tewit* in the production of a loud mellow whistle slurred upward in about the interval of a fifth. It may be whistled slowly thus:

About March, April, and the beginning of May, the skylark nearly always concludes its song, while descending from the sky, with a repeated plaintive whistle, which descends in pitch, and at this season is often uttered during the phrase. This cry is hardly to be distinguished from the call of the young skylark, and may be whistled slowly thus:

Later in the season the bird soars less when

singing, and generally omits this cry from his song, which then is rarely concluded with it.

I have heard a tree-pipit repeat in his phrases the squeak which is his call-note when he is migrating. Sometimes the chaffinch, when he begins to sing, in March or earlier, repeats the call-note he uttered when a nestling, and not that one he employed in autumn. He will repeat this note many times together, and thus produce the first part of his song.

The coarse call-note of the brown wren may be distinguished (though it is always uttered obscurely) in about one-half of the phrases sung in spring; it is sometimes uttered twice in one phrase. The goldfinch, house-sparrow — for undoubtedly the latter occasionally tries to sing — and the linnet appear to construct their songs wholly of call-notes and danger-cries. The blue tit, great tit, coal tit, and jackdaw have but little more originality. In the songs of these birds, and in those of the nightingale, swallow, grasshopper-warbler, stonechat, cirl, yellow, and common buntings, wood-wren, redstart, hedge-accentor, and others, are strong indications of the construction of phrases, wholly or partly, by the repetition of single cries.

Under the influence of habit, or of physical

strength, the phrase is sometimes modified in length and in force; and is often modulated in approximately the same successive intervals of pitch, so that a kind of musical strain is produced. The length of a phrase would obviously be dependent, firstly, upon the number of breaths necessary to its expression; thus, if a bird employed but one breath, as a cock does in crowing, the length of the phrase could not be great; but if respiration could be continued during the song, as it is during the extended phrases of the skylark, starling, grasshopper-warbler, and many others, the utterance would be limited only by fatigue of voice. It appears that many extended phrases are accompanied by a movement of the lungs similar to that which occurs in us during laughter; and this seems to induce a proportionately rapid movement of the wings, for when the notes are uttered slowly, as, *e.g.*, in the comparatively slow song of the willow-wren and the long notes of the wood-wren, the wings are very slightly expanded during or after each note, just as they are during each caw of the rook; but when the wood-wren sings its sibilous song, which is only a rapid repetition of one cry, the wings are shaken with increasing rapidity as the phrase proceeds to its ecstatic close, which is accompanied by a trembling of the whole of the body, and seems to be the very expression of passion. This movement of the wings in accordance with the frequency with which a single cry is repeated may be observed in the yellow bunting, cirl bunting, whitethroat, lesser whitethroat, and several other birds.

It is not surprising that bird-phrases should be repeated in the same intervals of pitch, or that the distinguishable notes which they include should follow the same orders of succession, as habitually occurs in all the more inferior singers; for amongst ourselves we detect a corresponding influence of habit in the wearying repetitions of street hawkers and railway porters. Just as loudness of voice may have been, and probably was, a factor in the survival of certain young, so a vehement and forceful exclamation of call-notes uttered by males to females may very possibly have been perpetuated, firstly, by the agency of selection, and, secondly, by the imitativeness of the young. The rivalry and emulation of the males, also, may have likewise instigated the introduction of modifications or additions to the phrases; and these, if agreeable to the females, would probably have been continued in succeeding generations by the agency of imitation.

It is not assumed that the statements in this chapter can be cited as proofs that all songs have been developed from call-notes ; for in some species, as in the domestic cock, the attempted song is more like the danger-cry than any call-note ; but I conceive that the evidence which I have brought forward is of value as indicating the history of the songs of many species of birds. It shows that the songs were, at first, mere repetitions of call-notes, or possibly of defiance-cries, which have since been more rapidly uttered and varied, with the result that novel strains have been slowly developed. The evidence adduced is but a small portion of that available ; and in this respect I have here, and indeed throughout the book, for the sake of brevity, instanced a few diverse species as representative of their respective genera.

CHAPTER VI

NOTICEABLE INCIDENTS CONNECTED WITH BIRD-SONG

1. SONGS are generally uttered by male birds only. Females which sing are generally mentioned by authors as exceptions to this rule: such are the female great grey shrike and the female Lapland bunting. The female starling sings: I have often observed a pair warbling together near their nest. Bechstein observed that nightingales, when courting each other, "utter a gentle subdued warbling." Gilbert White noticed that while the quack of the common duck is loud and sonorous, the voice of the drake is scarcely discernible (*Nat. Hist. Selborne*, p. 116). Rennie recorded the singing of females of the following: redstart, blackbird, willow-warbler, and bullfinch (*Domestic Hab. Birds*, p. 262); and he observed that the songs of the females are

like the recording or recovering of voice by the males. In this connection may be mentioned some duets noticed by Mr. Hudson in his *Naturalist in La Plata*. Describing certain South American treecreepers, he says, " On meeting, the male and female, standing close together and facing each other, utter their clear ringing concert, one emitting loud single measured notes, while the notes of its fellow are rapid rhythmical triplets ; their voices have a joyous character, and seem to accord, thus producing a kind of harmony" (*op. cit.* p. 256). In one species these duets are practised by the young while yet in the nest. Mr. Hudson says of the South American tyrant fly-catcher that in a majority of cases their songs are simply joyous excited duets between male and female. The songs of the woodhewers, or *Dendrocolaptidæ*, also are chiefly duets. " In the red oven-bird (*Furnarius*) the duet consists of triplets uttered by the male, with a strong accent on the first note, while the female keeps even time with single notes. The *finale* of this performance consists of three or four notes uttered by the second bird alone, in an ascending scale, the last very piercing" (*op. cit.*).

Darwin considered bird-song a charm or a call-note addressed to the female (*Descent of*

Man, p. 368), and such it probably is during the breeding season; but the first songs of immature birds, such as the young skylark, robin, and thrush, cannot reasonably be considered to be directly occasioned by the emotion of love.

2. Song is not uttered until birds have nearly attained their full size. Mr. W. H. Hudson asserts (*op. cit.*) that the young of the red oven-bird when only partially fledged sing the song of their parents; this is unlike the hunger-cry, which resembles that of other fledgelings. The Rev. H. A. Macpherson informs me that bullfinches try to pipe as soon as they can perch, and that young hawfinches when in confinement sing almost as early.

3. Birds sing most frequently at early morning or at evening, and in the former case they commence singing in much less light than that which remains when they cease in the evening. They also sing in warm, fine weather, rather than during cold or fog. But indifference to weather is by no means infrequent, for many birds sing much during rain—for example, the blackbird, robin, willow-warbler, and chiffchaff—while house-sparrows are particularly noisy during rain, especially in winter, and when they are in the shelter of a tree. The sparrows near my bedroom—a numerous population

—chirp noisily every morning except when frost is severe; and then, if the cold is intense, not a single chirp can I hear; but directly a thaw comes, all the birds at once announce the change of temperature. The nightingale, grasshopper-warbler, sedge-warbler, and landrail may be mentioned as singing at unusual hours.

The stimulus which induces birds to sing particularly at dawn may be the same as that which leads them to the same exercise in the rain—the pleasure of anticipating approaching incidents: in the one case, daybreak, which brings comparative safety to the waking creatures; and in the other, the coming forth of many kinds of insects and larvæ. The insect-eaters, at least, would be able to appreciate the latter event, and that may possibly account for their singing more than the finches during rain. There is nothing to show that erotic emotions are in any way excited by dawn or by rain. The habit of singing early and late may also have been encouraged by the enforced leisure which many birds must then undergo, when they cannot seek food, and when their young need no sustenance. The nightingale's habit of nocturnal singing is easily understood: the male sings to attract the migrating female, and does not seek her. She flies by night, as is the habit of the male; the latter, therefore, has very naturally developed the habit of singing at night, not only when looking for the advent of a mate, but even after she has arrived. Bechstein observed that the nightingales of some districts sang by day, and those of others by night, a local variation which has been several times recorded. Mr. Hudson states that, in a village within an hour of London, the nightingales sing only by day (*Birds in a Village*, p. 8).

Yarrell observed that whenever at evening the robin perches high, and begins to sing merrily, the following day will be fair (*Brit. Birds*, vol. iv. p. 308); but I cannot say that I have noticed this as a definite incident, for in autumn the robin habitually sings after a storm, and from an elevated position. The effect of rain is observable in the cries of some birds which, like the house-sparrow, cannot be called singers. Wilson observed that ospreys (*Pandion haliaëtus*) "gambol in the air before a change of weather to rain" (*Am. Orn.*, vol. ii. p. 115); he also remarked that the great northern diver (*Colymbus glacialis*) calls much before rain (*ibid.* vol. iii. p. 109); and that a woodpecker (*Picus pileatus*) near rivers is noted for "making a loud and almost incessant

cackling before wet weather" (*ibid.* vol. ii. p. 20).
Yarrell has stated that the term "storm cock" was
applied to the mistle-thrush because of that bird
"giving his song both before and during the occurrence of wind and rain." It appears to me, however,
that this thrush sings most during bright sunshine,
especially in the first three months of the year.
The domestic cock readily perceives an approach
of rain, and announces the fact by crowing.

Dense fog is a powerful deterrent of song: I
have several times witnessed its almost immediate
effect in silencing singers. On one occasion this
was particularly noticeable. One day in January
several starlings, robins, and a brown wren were in
full song in a valley, when suddenly a low cloud swept
up the valley, shutting out the sunshine, and rendering indistinct or invisible the hedgerow elms. The
songs abruptly ceased. After the lapse of some
ten minutes the cloud passed away, the sun suddenly
shone, and all the birds recommenced their songs.
I had observed carefully only a few of these,
when another cloud obscured the scene as quickly
as its precursor, and at once the songs were abandoned. I waited during a longer period of misti-
ness, and then turned homeward; yet I had not
gone beyond earshot of the place when the second

cloud was swept away, and I heard the birds once more resume their songs.

There is no reason why we should not credit birds with the possession of a sense of pleasure in the aspect of their surroundings. Sterland seems to have had some such idea when he wrote that the song of the robin in autumn "lacks the joyousness of spring, and the bird, in sympathy with the departing season, seems to breathe a plaintive and melancholy strain" (*Birds of Sherwood Forest*, p. 63). At that period, especially on still days, the song is often preceded by the utterance of a slow succession of sad-sounding repetitions of the call-squeak. "Many species which become silent about midsummer resume their notes in September (or October), as do the thrush, blackbird, woodlark, willow-wren, etc. Are birds induced to sing again because the temperament of autumn resembles that of spring?" (White's *Selborne*). These autumn songs may be induced partly by an inclination towards love, although both robin and starling often sing in August when quite alone; the former bird, however, in song obviously challenges others to fight. It appears that only the young larks, thrushes, and blackbirds of the year sing in September or October.

In suggesting that some wild birds may derive pleasure from the appearance of their surroundings, I do but put them on a par with the homing pigeon. This bird, in order to return to a former abode, will abandon both mate and young. The male pigeons will "home" better when their mates are sitting than at other periods. It may be as well to remark here that the homeward flight of these pigeons is not directed by any blind instinct, but by sight and memory; for the birds cannot find their way through thick fog, even from the distance of a few hundred yards. I have excellent authority for these statements. It has been noticed in many birds that the same sites for nests are adopted year after year.

In this connection may be mentioned the "instrumental music" which many writers have observed, and which may be defined as consisting of sounds other than vocal sounds uttered by birds, as, for instance, the sounds caused by the smiting together of wings, drumming sounds produced by rapid blows of the beak, and similar noises. And in conjunction with this feature we may notice the strange and seemingly purposeless dancing and other antics by which song may be accompanied. Mr. Hudson writes: "There are human dances in which one person performs at a time, the rest of the company looking on; and some birds in widely separated genera have dances of the same kind. A striking example is the *Rupicola*, or cock of the rock, of tropical South America. A mossy, level spot of earth surrounded by bushes is selected for a dancing-place, and kept well cleared of sticks and stones. Around this area the birds assemble, when a cock bird, with vivid orange-coloured crest and plumage, steps into it, and, with spreading wings and tail, begins a series of movements as if dancing a minuet; finally, carried away with excitement, he leaps and gyrates in the most astonishing manner, until, becoming exhausted, he retires and another bird takes his place."

4. Many phrases sung by the better singers tend to rise in pitch towards the end. There are abundant exceptions to this rule—for example, the songs of the chaffinch, willow-wren, greenfinch, green woodpecker, and others—but it prevails in songs of the wood-warbler, nightingale, blackcap, golden-crested wren, blackbird, and yellow bunting.

5. Phrases are uttered with increasing vehemence towards the close, as though the emotion which caused them became intensified during expression. In April the starling, with flapping wings, illustrates

this theory. With the exception of the monotonous songs of the nightjar, grasshopper-warbler, and cirl-bunting, there are few which do not exhibit this feature. Even the stock-dove's jerky *coo* evidently becomes more vehement as his song proceeds.

6. Individual variation is generally to be perceived, if at all, at the end of the phrase; and similarly, the first parts of the phrases sung by allied birds have most resemblance to common types. This may be observed in the whitethroat, lesser whitethroat, blackcap, greenfinch, brown linnet, lesser redpoll, siskin, sedge-warbler, reed-warbler, and in the buntings.

7. The alarm-cries and call-notes (which are generally common both to males and females) of allied birds are more alike than are their songs—which are almost invariably exhibited by males only.

8. The cries of allied young are more alike than are the cries of allied adults. This, and the preceding subject, will be considered in relation to divergence from the general to the special as exemplified in the voices of birds, and in the chapter on heredity.

9. Only small birds properly sing. This is the statement of Darwin, who mentioned the Australian

Menura Alberti as a notable exception to the rule; for the bird, "which is about the size of a half-grown turkey, not only mocks other birds, but its own whistle is exceedingly beautiful and varied" (*Descent of Man*, p. 371). The crested screamer (*Chauna chavaria*) is another exception.

Why are singing birds small? Is it because certain female birds, in choosing their mates, have always neglected the attractions of size and strength in favour of music? Or is it because the larger birds of a species are more attractive to the kinds which prey on them, and hence have become silent through fear? Or, again, are small individuals naturally more vivacious than others of superior size? Certain it is that the largest birds of various races are by no means good singers (*e.g.* raven, mistle-thrush, hawfinch, cushat or ringdove, mute swan, eagle), and that some of the smallest of their allies are more musical (*e.g.* jackdaw, nightingale, linnet, turtle-dove, duck, and chanting falcon—which is known to sing somewhat fluently). Probably all these influences have combined to render small but musical individuals of various species successful in the contest for mates. It is also probable that mode of life has affected the development of voice as well as mere size. Darwin says that "all the common songsters belong to the order of Insessores (or perchers), and their vocal organs are much more complex than those of other birds; yet it is a singular fact that some of the Insessores, such as ravens, crows, and magpies, possess the proper apparatus (Bishop, in Todd's *Cyclop. of Anat. and Phys.*, vol. iv. p. 1496), though they never sing, and do not naturally modulate their voices to any great extent" (*Descent of Man*, p. 370). It is probable that the Insessores, and vast numbers of pre-existing species, have always been of arboreal habits, and have thus been dependent on the voice for a means of intercommunication when at some little distance apart, and especially for a means of announcing the approach of an enemy through the thick foliage in which so many of the insectivores spend most of their time. Consequently their voices, and simultaneously their hearing, would have been gradually developed; and the latter feature, from having been so continually exercised in the vital necessity of detecting signals of danger, or mere cries of distress, would have become at once delicate, critical, and accurate, both in males and in females; and the latter would therefore have been competent to detect, and perhaps liable to have been attracted by, any abnormal powers of melodious utterance in their suitors.

The conjunctive abnormal development of the organs of hearing and of the voice in birds does not prove that the ear was evolved earlier than the voice. However, we know that many birds with acute hearing do not sing much, and also that some animals which have ears have no voices—such are our British lizards, at least—it is therefore probable that the ear was developed prior to the voice, and that its powers were greatly increased before the voice became so strong as it is in some of our birds. For the voice is only occasionally employed even by some of the best singers, which are always acutely sensitive to any alarming sounds.

The present evident necessity for the frequent use of the voice by arboreal birds is proved by the noisiness of flocks of them, and by that of family associations of a few individuals, such as those of the titmice, nuthatch, etc. Birds which have hardly more than one cry are compelled to make abundant use of it, if they live in the obscurity of trees: such are the woodpeckers. It is practically as true to say that all the arboreal birds are small, as to say that all the singing-birds are small; and it is evident that a life in trees is as conducive to smallness as it appears to be to a development of the sense of hearing. Leaf-eating and bark-boring grubs and insects are not sufficiently abundant to support any but small creatures; and most of those animals which subsist upon this food have to work hard for it, at least during the greater part of the year. Also, small birds would have an advantage over larger ones in the pursuit of insects among dense foliage. Thus the trees can support a numerous population of small birds, though offering but meagre fare to larger kinds; and the birds which dwell among the trees have not only less need of caution in the utterance of cries than have birds on open land, which are visible from a distance, but their sense of hearing is continually exercised, as we have seen.

The ravens, crows, and magpies mentioned by Darwin as being not given to song, are very observant of sounds; and many of them are garrulous, though, on account of their timidity, their loquacity is not readily observed by us when they are in a wild state. If these larger birds, or their remoter ancestors, had been compelled by their size to abandon arboreal life for the open land, or had increased in size in consequence of such abandonment, they might have retained the apparatus (which they actually possess) for singing, even when their more exposed position rendered song an accomplishment too dangerous to be continued.

10. Accent in song. Many species, and especially the brown wren, hedge-accentor, thrush, and blackcap, accentuate certain notes in their songs, thus producing a rhythmical progression throughout the phrase. The blackbird often exhibits the same feature.

11. Birds which sing well and continuously are generally clad in sober hues; and the converse is equally true, namely, that brightly coloured birds are not singers. There are, of course, some well-known exceptions—such as the American bluebird, our own goldfinch, and the great titmouse. If, as appears to be the case, brightly coloured birds and singing-birds have been developed from common types, there must have been some cause for this divorcement of colour from song in the avian races. Is it not probable that, just as some female birds may have been influenced by the songs of their suitors, so others may have been similarly affected by the brightness of hues which certain males may have displayed; and also that as arboreal birds may have developed a great sensitiveness of hearing by listening to the notes of others hidden from them by leaves, so other birds, from specially exercising the eyes for a similar purpose, may have developed a habit of scrutinising every object very intently, and thus have been induced, all unconsciously, to prefer

mates who appealed to their sense of sight rather than to that of hearing? But, it may be asked, do birds which live in trees ever depend upon their eyes rather than upon their hearing for the detection of an enemy? In our temperate climate there are practically no enemies whose approach through trees will not be more readily perceived by ear than by eye; but in tropical climates there are other foes to reckon with—a numerous race of tree serpents, whose approach is silent, yet terribly dangerous, and whose colours often resemble the general tone of their surroundings. These animals are great enemies of birds, who must watch carefully for them, as well as listen for sounds announcing the arrival of other predacious creatures. If such watchfulness would have had the effect of developing and rendering critical the powers of sight, it may have induced birds to prefer mates whose hues appealed particularly to this sense; and it would thus explain why arboreal birds which live in tropical climates are often so brightly coloured. But the arboreal birds in temperate climates have few or none of these serpent enemies to fear; hence they depend upon the ear rather than upon the eye for the detection of an enemy, and when hearing no alarm they may be at ease.

12. The primary necessity to the development of varied song in species or individual is leisure. One cause of the persistence of the songs of caged birds may be found in this result of their changed condition of life — that they have nothing to do but to sing. The wild bird has always plenty to notice and consider,—the approach of various creatures—men, beasts, hawks, and other birds; the sounds which these produce, and which signify various degrees of safety, or of peril; the indications of food in air, or tree, or on the ground; and, lastly, the state of the atmosphere, and the various weather-signs which all birds observe;—such incidents as these occupy the wakeful hours of the wild bird. But the caged bird—often secluded from all communication with his kind (one, perchance, of a gregarious species), without the necessity of seeking food, with a horizon limited perhaps by a smoky garden, perhaps by a dingy window—can take no exercise but in hopping from perch to perch, across and across his cage; and can hear no call-notes but his own, which he repeats again and again, and, if he has been reared in a cage, his own song, which he seems to utter as much for the sake of such occupation as it affords, as to express by means of it any desire for a mate, or any pleasure in his

surroundings. Rennie could suggest no key to the diversity of learning in captive birds, especially in the nightingale, which will learn the notes of other birds and retain them after it has heard its own species—"unless it be that, from want of other amusement, it (the captive) listens more when it is confined" (*Domestic Habits of Birds*, p. 278).

Mr. Hudson raises his voice against what he considers a pestilent delusion, namely, that all wild animals exist in constant fear of an attack from numerous enemies watching for an opportunity to spring upon and destroy them. He says that the truth is, that they are free from apprehension unless in the actual presence of danger. "Suspicious at times they may be—but the emotion is so slight, the action so almost automatic, that the singing-bird will fly to another bush a dozen yards away and at once resume his interrupted song" (*Birds in a Village*, p. 8). The demeanour of wild birds seems to me not to bear out this view. Except in the season of love—with many creatures a period of indiscretion—they are all keenly watchful for the approach of enemies. Such birds as the woodpeckers and titmice, when bark-searching, or others feeding on open ground, are very persistent in diverting their attention from their expected food, in order to discover at the earliest opportunity the approach of an enemy.

There are times when birds which sing practically throughout the year, such as the starling and robin, abandon their songs. Such an occasion is observable when their young are nearly ready to quit the nest. Then these birds will sing at early morning and late evening, before and after the labour of the day—the supplying of the wants of their young—has been performed; but when the young leave the nest, and parents are occupied all day in obtaining food for them, the songs are quite abandoned, from the very simple cause that the birds have no time for singing. How can the robin, hunting everywhere for insects and worms, waste a moment when his expectant young demand his constant care; or how would it be possible for a starling, which, at the same period, may often be seen panting from fatigue (especially on a hot, dry day), to sing his usually long and noisy phrases? These birds, and others in like circumstances, do not then sing, because they are either seeking food for their young, or are intently watching lest enemies approach. In either event they have not the necessary leisure in which to practise song. I suggest that a want of leisure may have been a potent cause of the lack of individual variation of phrases in the birds of the tropical regions, in which localities a redundant population of birds is kept in check by the attacks of enemies, or by mutual contests, rather than by changes of climate, by which in temperate lands it is periodically reduced. We must remember that in the latter places the arrival of warm weather calls forth sustenance not only sufficient for the birds which have wintered there, but also enough for swarms of immigrants who crowd to the feast, and even rear their young with this superabundant food. When the winter in a temperate land is severe, so greatly are the food supplies lessened that vast numbers of hardy birds perish by starvation, though they seek never so persistently for food; but directly the weather becomes genial, the creatures upon which they subsist spring into life on every side, and then some of the birds can spend long hours in doing nothing, or in singing. This increased leisure is a first necessity to the singing of our native species; and I submit that the want of it may fairly be suggested as a cause of the lack of varied song not only amongst tropical birds, but also in those which in any part of the earth obtain their food in a particularly laborious manner—such as the woodpeckers and creepers, which are practically always at work.

13. Birds which are habitually laborious, watchful, stealthy, or accustomed to gorge themselves with food, are habitually silent. Several birds exhibit all these characteristics. Among watchful birds may be mentioned the kestrel, butcher-bird, and others of similar habits, which watch for prey ; also rasorial birds, most of which live or feed on open land, and for this reason are compelled to watch for the arrival of their enemies, many of which, approaching in flight, may be seen at a great distance. Birds stealthy in movement are generally those which attempt to take their prey unaware— such as the hawks—or which seek to escape unobserved from their enemies ; of the latter the pheasant is a notable example. Birds preying on animals of comparatively large size generally gorge themselves at every meal ; and in this condition they become inert, and apparently are equally passive in mind, uttering hardly a sound while digesting their heavy burden. Of this kind are the divers generally, and also the larger predacious birds. Many of these last are alternately watchful, stealthy, and lethargic, as they seek, attack, or digest their prey. Song is foreign to their natures, but their usually loud shouts, screams, or calls, are employed as signals when they have wandered far apart in the search for food.

14. *Flight in song.*—It seems to me that flight is not resorted to during song merely to enable the singer to utter his notes from an elevation, but that in several instances the manner of the flight itself is intended to suggest to a female bird the attention of a suitor; and that when so employed it is to some extent analogous to that flapping of the wings which gives expression to the wants of many young birds when about to be fed.

The greenfinch sometimes exhibits, perhaps, the most characteristic flight when singing, its wings being fanned slowly, and with their widest sweep. The sedge-warbler sometimes behaves in a similar manner. The wood-warbler flies in song with the same stroke of wing, as I have seen it, at the distance of eight or ten feet. Individuals differ in the extent to which this habit is followed. One at Leigh woods flew during twenty-seven out of thirty repetitions of the sibilous phrase. The legs of the wood-warbler are extended during this performance, like those of a rook just after it leaves the ground; in the former bird this attitude probably permits of a slower flight, just as in the latter it undoubtedly favours the execution of a more vertical ascent. I have seen a blackbird dart from his perch, in the middle of a phrase, and gesticulate while singing on the wing, and descend with a sweeping flight on wide-extended wings. It is possible that songs are uttered more easily when the wings are moved slowly; but however this may be, there is no doubt that many male birds sing while on the wing. Chats, warblers, finches, buntings, and pipits have been observed to fly when singing. Some birds, such as the common bunting and wood-warbler, habitually shiver the wings when singing rapid notes; and this suggests that the shiver referred to is caused by the movements of the lungs; as already stated, these, in many phrases, seem to be similar to those which occur during human laughter. Mr. Hudson describes the extraordinary song-flight of the crested screamer, a bird of about the size of our heron. He says: "The chakars, like the skylark, love to soar upwards when singing, and at such times, when they have risen till their dark bodies appear like floating specks on the blue sky, or until they disappear from sight altogether, the notes become wonderfully etherealised by distance to a soft silvery sound, and it is then very delightful to listen to them. The chakar, . . . so ponderous a fowl, leaves its grass plot and soars purely for recreation, taking so much pleasure in its aërial exercises that in bright warm weather in winter and spring it spends a great part of the day in the upper regions of the air."

Many birds give a fluttering or shaking of the wings as a means of address to their mates; such are the starling, female house-sparrow, greenfinch, and very many other birds which authors have noticed in all parts of the world. The young of many, if not of most species flutter their wings when being fed by their parents; and it is a common thing for vehement singers to flap their wings or, like the redstart, shake the feathers in their tails when singing. The lark and its near ally, the tree-pipit, fly much more during song in early spring, when they are courting, than in June and July, when their young are out of the nest.

15. Ventriloquism. — In regions where the enemies of birds are numerous, individuals having the power to utter call-notes in such a manner that the exact spot whence the sounds proceeded could not be determined by enemies, would probably enjoy a consequent advantage in the battle of life, and would not necessarily be prejudiced by this circumstance when seeking mates, who would be discovered as soon as they had alighted in the vicinity of the exclaimers. Certain it is that several good observers of bird-notes have been deceived by the

so-called ventriloquial powers of some kinds of birds; and it is possible that the lower animals might make the same mistake.

16. Singing in chorus.—Observers have recorded the simultaneous singing of many birds of the same species when associated in a flock—an incident by no means limited to one kind of birds. Wilson relates that this habit has been noticed in the following American birds: the rice troupial (*D. oryzivorus*), the red-winged starling, and the goldfinch (*F. tristis*). It may be observed in the common British starling and linnet. The screaming of the adult swift (*Cypselus apus*) is occasionally uttered in chorus, not more than once or twice by a solitary swift, and sometimes continuously by a number flying together. I have seen seven male swallows singing together perched on telegraph wires.

The starling appears to sing in chorus unintentionally, from the mere fact of several singers being together, but on two occasions I have known several starlings join in concluding a certain musical phrase commenced by one of their number—a performance which was repeated so many times consecutively as to be remarkable. Directly one of them began the phrase with the following notes :—

the others instantly joined in with the concluding notes:

The whole phrase was

This incident occurred in the churchyard at Bisley, Gloucestershire, and was repeated there on several days when I visited the spot. There were about six starlings singing. The intervals were fairly correct, and the unison was seemingly perfect; from which we may infer that the phrase had been much practised by the birds, having possibly been originally learned by one of them from some captive, or from the church music, which during eight centuries or more has been heard in that place. I also heard the same song uttered in the same way by starlings at the Conegre, three miles from Bisley, and in the following year I again heard it at the same places; I have never heard it elsewhere. This was the only instance of what may be called intentional chorus-singing that I have observed. Mr. Hudson, however, has recorded a wonderful occurrence of it in the crested screamer of La Plata:—" Travelling alone one summer day, I came at noon to a lake on the Pampas, called Kakel—a sheet of water narrow enough for one to see across. Chakars in countless numbers were gathered along its shores; but they were all arranged in well-defined flocks, averaging about five hundred birds in each flock. These flocks seemed to extend all round the lake, and had probably been driven by the drought from all the plains around to this spot. Presently one flock near me began singing, and continued their powerful chant for three or four minutes; when they ceased the next flock took up the strain, and after it the next, and so on, until the notes of the flocks on the opposite shore came floating strong and clear across the water, then passed away, growing fainter and fainter, until once more the sound approached me, travelling round to my side again. The effect was very curious, and I was astonished at the orderly way in which each flock awaited its turn after the first flock had given the signal" (*op. cit.* p. 227).

CHAPTER VII

THE INFLUENCE OF HEREDITY IN THE PERPETUATION OF THE CRIES OF BIRDS

IT is clear that birds inherit the desire and power to sing, and style of song, and that some definite cries are perpetuated solely by the same agency. The young of the common fowl, turkey, pheasant, partridge, duck, and goose (at least) inherit their cries; and, whether reared naturally or artificially, are equally willing and able to employ them upon the appropriate occasions. But, so far as regards common chickens, this observation is not in accord with the opinion of Professor Hill-Tout (Principal of Buckland College, Vancouver), who writes to me as follows :—

"I have noticed that chicks reared by hand present many interesting differences from your point of view from those bred entirely by a hen. With the mother, they show signs of fear of human beings much sooner than those which never had a mother hen to look after them. This I judged to arise from the fussiness and anxiety of the mother in the early days of their life. She sounds her warning and danger-notes so frequently that the emotions of the chicks are perpetually stirred; and they, too, become fearful and suspicious, and frequently sound an alarm without any apparent reason. With hand-bred chicks this is the reverse, and the alarm and danger-notes are uttered much later in life and very much less often; and the alarm-notes of other fowls do not seem to affect them nearly so much as they do chicks reared by a hen. The conclusion I came to on this head was, that though the sounds emitted by chicks are undoubtedly instinctive, and, at first, of little or no value as warning cries to others, they do soon acquire a suggestive value of danger, in the minds of other chicks, but that this emotional import is more the result of association and experience than something inherited."

Chicks hatched in incubators, however, will utter distinct alarm-notes, even when but a day old. The Rev. H. A. Macpherson, whose contributions to ornithological science are well known, writes of some young golden plover, hatched indoors, that as soon as they could run piped like the old birds, whose note they had never heard (*in litt.*). The Rev. A. R. Winnington-Ingram, a careful observer, informed me that he had heard the cry of the lapwing uttered by young birds of that species while in their shells. There could have been no mistake about the matter, for the eggs containing the birds had then been removed to the rectory at Lassington. Birds also may inherit a knowledge of the cries of their parents, and some of them when only lately hatched, or, according to Mr. Hudson, when chipping their way out of the shells, immediately become silent if the parent utters a cry of alarm in the vicinity.

The latter observer was of the opinion that some young of the rhea, or South American ostrich, reared by him, had so strongly inherited this knowledge, that, when he imitated the alarm-cry of the species, they ran to him for shelter. But it appears that the birds in question had been captured "just after they had hatched out," and therefore they might possibly have been partly educated by their parent, for we know that very young birds are generally apt pupils. Mr. Hudson also records similar instances of inheritance in a South American tinamu and the ovenbird, the young of which, when only a day or two old, utter the full cries or songs of their respective parents (*Naturalist in La Plata*). Apparently, birds which inherit their cries never imitate the notes of other birds. It is probable that the birds which now inherit cries have always done so, and that the notes have been partly developed by the agency of selection, acting on slight and accidental modulations; for, despite the force of inheritance, variations are of frequent occurrence.

It might be well here to notice that there are some species of song-birds in which both cries and songs are perpetuated, not by heredity, but solely by imitation, though whether the mimicry is voluntary or involuntary would be difficult to prove.

It will be observed that the young of certain species, such as the fowl, pheasant, partridge, turkey, swan, duck, cuckoo, etc., though inheriting the notes of their parents, do not at first utter all of them, but gradually acquire them in course of progress towards maturity. This feature may be consequent upon physical development; but the young of the golden plover, lapwing, tinamu, and oven-bird possess, as we have seen, the full notes of their parents. This divergence of habit cannot be attributed to any difference in the relative size of the species, for the tinamu is nearly as large as a fowl; but it indicates that the cries most fully uttered by the young have been inherited during the longest period.

We know that certain birds inherit their cries. I propose now to call attention to certain sounds which may be called "family cries," because they are common to species which exhibit physical affinities, and have therefore been classified as belonging to the same families or groups of birds. In dealing with this subject, I do not purpose to distinguish notes perpetuated by inheritance from those perpetuated by mimicry. If the young of some song-birds acquire every danger-cry and call-note solely by imitating the voices of their parents, and keep strictly to these models, and the notes are common to all birds of the same species respectively, a similarity between those notes in allied species is just as valuable an indication of a common origin as it would have been if the perpetuation of the notes had resulted from inheritance. For, in some species, filial mimicry of voice seems to be as powerful as inheritance of form and colour; but this occurs only in singing birds or in their allies.

In this chapter it will be necessary to mention many specific and other cries, in such a manner as to convey a definite idea of the sounds under consideration. I know how difficult of realisation such attempts must be; yet I hope that by means of descriptions of the occasions, and of the manners of birds, when such cries are uttered, in conjunction with a careful suggestion of the tones employed, the reader will be enabled to understand my meaning, even if he have not previously been familiar with the actual cries indicated. I feel sure that if any one, able to distinguish the commoner British birds, will take a stroll in the country, provided with a small telescope, or a powerful pair of field-glasses, and will notice all the cries of birds heard during that single excursion, especially if he be accompanied by a dog, he will hear much illustrative of my remarks. It is advisable that a dog should be his only companion; for if a friend be near, conversation will arise, with the result that the notes of birds will not be heard, and in any case the birds will be scared away sooner than if disturbed by only one person. The dog, running among wayside bushes, will occasion the utterance of many interesting cries of alarm.

The study of this subject enables us to perceive with increasing clearness distinctions which are correspondingly subtle: this is a truism which needs no demonstration; and it applies with equal force to the observation of bird-song, or to that of any other matter. An observer can educate his memory to the recognition of a multitude of cries, and even of vocal inflections, which are common to different species.

It is notorious that animals which have similar physical features are much more alike in the early stages of development than when mature; in short, that in progressing towards maturity the young tend to diverge from general types to special types. The same progressive divergence is indicated in the voices of most genera of birds. In fact, the cries of the young of birds physically allied are more alike than are the cries or the songs of adults of the same species. The reader who wishes to prove this theory will be enabled to do so by the following records of family resemblance, in which danger-cries and callnotes are the principal features discussed. I shall therefore not deal with this theory by itself; and the omission will avoid an extensive repetition of instances which are mentioned in the following pages.

The similarity between the cries of some few species unrelated by physical resemblances (such as the *pink, pink* cry of the great titmouse, which, as Mr. A. Holte Macpherson informs me, is common with that bird in a part of Scotland—as it is in Kent and in Gloucestershire—and which is closely like the *pink, pink* of the chaffinch) afford instances of exception to the almost universal rule that the cries of birds nearly related physically are more alike than those of others which cannot be thus associated.

The *Corvidæ*, or crow family, do not present such obvious vocal resemblances as do some other races of birds mentioned in subsequent pages. Yet it may be well to commence this theme with an order in which a certain broad tone is widely prevalent.

Birds of the crow family, as a rule, have little or no song; but some members of the family possess considerable power of mimicry, though they do not in a wild state utter sweet tones. The raven may be placed at the head of the list, as showing in a feral state the least indication of musical ability; and this absence of song is not strange in a bird which, however imitative, lives in remote and infertile places. The carrion crow utters the same kind of cry as the raven, but at a higher pitch, and, like that bird, is capable of being educated to the pronunciation of words. The crow common in British Columbia is smaller than *Corvus corone*, and utters a similar but less harsh cry. The hooded crow is said to have a "rather harsh sound resembling the syllable *craa*" (Macgillivray, *Brit. Birds*, vol. i. p. 532); and the authority who records it compares the cry of the rook (*ibid.* p. 544) to the syllable *khraa*; hence we may conclude that the two cries are of much the same intonation. The rook, when greatly alarmed, utters a low croak; so does the jackdaw when alarmed on account of its young. The jay employs a high *cah*, modulated, on the like occasion. The blue jay of America (*Cyanocitta cristat*) is described by Wilson as being a great mimic, and as uttering cries like those of our jay, when alarmed. The chough imitates the human voice well (4th ed. Yarrell, vol. ii. pp. 15, 16). The nutcracker utters as an ordinary note *crah crah crah*, or *cru cru cru ;* and Mr. Warde Fowler has seen it "croaking like a small raven" (*Summer Studies of Birds and Books*, 1895). The starling utters as a vehement alarm, and especially when its young are in danger, a cry *cah*, closely like that of the rook and jackdaw, and both sexes employ it. Young starlings in the nest utter cries closely like, but less loud than those of rooks of the same age. The starling has also a note which seems to relate the species to the thrushes: I allude to the high, squealing, and almost toneless notes which, in the breeding season, nearly always conclude its song-phrases, are sometimes evidently uttered as a vehement call to the female, and are very frequently accompanied by a flapping of the wings. These notes are rarely uttered in the fall of the year: in the vernal season they are most persistent. I do not know of this cry being uttered by any other of the *Corvidæ;* but the redwing thrush, when migrating, utters just such a toneless, though a much shorter squealing cry or call-note. The inferences deducible from this resemblance are strengthened by a physical analogy between the starling and the *Turdidæ*, the fact that the young of the former are brown when first fledged. Some of them, having retained their nestling-plumage until maturity, have been held to constitute a distinct species, the solitary thrush of Montagu, Bewick, and Knapp (4th ed. Yarrell's *British Birds*, vol. ii. p. 242). Not only does one of the *Corvidæ* utter a cry like one of the *Turdidæ*, but one of the latter appears to utter a cry resembling the most characteristic tone of the former. I have heard the mistle-thrush, when carrying food, apparently to its young, utter a loud *cah*, like the cry of the jay.

In dealing with birds of the thrush family, it may be well to treat first of their alarm-cries and their call-notes. The loud, rattling alarm-screech of the mistle-thrush is uttered by its young when able to fly, as a call to the parents; it is uttered most often and most vehemently by the adults during spring, and is somewhat abbreviated in winter, when it is, apparently, equally a call and an alarm. In early spring this cry seems to be sometimes uttered, as is occasionally the rattling cry of the blackbird, for the mere purpose of making a noise—probably to attract attention. It is heard in a slightly abbreviated form as an alarm uttered by the common thrush when the nest or the young are threatened by a predacious beast or bird; but, apparently, upon no other occasion. Mr. Warde Fowler considers that the alarms of blackbird and of ring-ousel are generically similar, but specifically distinct (*in litt.*). Mr. Harting says of this species: "When on the wing, its note has been compared to the noise made by striking two large stones together" (*Birds of Middlesex*, p. 36). Bechstein wrote that the bird calls *tak* like a blackbird (*Nat. Hist. Cage Birds*, p. 201). The fieldfare utters a somewhat similar alarm and call-note, which Mr. Harting renders *tcha-cha-cha* (*Birds of Middlesex*, p. 30). The blackbird's rattling alarm, more musical than that of the mistle-thrush, is familiar to all dwellers in the country. It is much modulated in different individuals. Layard says of the South African olivaceous thrush (*Turdus olivaceus*), when alarmed and hurrying off, "its voice and manners reminding an observer of the European blackbird" (*Birds of South Africa*, p. 201). The strong family resemblance between the rattling alarm-cries of the mistle-thrush, thrush (when defending its young), ring-ousel, blackbird, and fieldfare is easily traced. I heard the cry in question uttered by the American robin with much of the manner of the blackbird; but in the former it is the less prolonged. I have observed this bird in Ottawa and in British Columbia. In size and colour the bird is akin to a hen blackbird. In general habits it resembles that species. So far, I have never heard an analogous cry uttered by the redwing; but then, the common thrush never utters this cry in winter when the redwing is with us; and it is very possible that the latter, when defending its young, reverts, like so many other birds, to some cry prevalent in the genus to which it belongs, and therefore probably of remote descent.

I would draw attention to the great similarity between the mode and occasion of deliverance of the rattling alarm of the blackbird and that of the robin, the only differences being that the alarm of the robin is uttered rather less often during flight, is less loud, and less varied in pitch. The redstart—whose young, like those of blackbird and of robin, are at first of a mottled brown hue on the breast—utters as part of an alarm to its young a sharp snap, sounding like the noise made by striking two small pebbles against each other; when the bird is greatly alarmed it repeats this note many times in succession, and thus produces a rattle not unlike the rattling alarm of the robin. The young robin, when about to be fed by its parent, frequently repeats its call-squeak, which is merely an immature utterance of the call-squeak of the adult; and when food is being given to it by the parent, a curious repeated sound, like the word *cah* whispered very loudly, is also heard. When the young redstart is fed these two sounds are heard, in tones so like to those of the robins, that I have several times mistaken the identity of the birds which uttered them. The telescope enabled me to arrive at a correct conclusion on the matter. In the adult redstart the sharp click of alarm is generally preceded by the single utterance of a short upwardly-slurred whistle, somewhat louder than, but otherwise like, the alarm and call-note of the young and adult chiffchaff and the full-grown and adult willow-warbler. It may be whistled thus:

The wide range of this cry, which I have heard uttered by redstarts at Stroud, Dean Forest, New Forest, Oxford, and other places, and the fact that it is an alarm-cry, are of value as linking the redstart to the true warblers. Whether or not the snapping note of the redstart resembles the short note, *tac tac*, of the blue-throated warbler (*Ruticilla suecica*, Linn.) (Bechstein, *Cage Birds*, p. 235), or the "short snapping note of the latter" (1st ed. Yarrell, vol. i. p. 236), I cannot say; but I know that the *tack* alarm of the blackcap is not unlike the *chick* alarm of the redstart. Yarrell observed that when bluethroats are alarmed by any one approaching their nest, their notes of alarm or anger resemble those of the nightingale. He also noticed that the bird very frequently jerks up its tail in the manner of the robin and nightingale (*ibid.*). The *tack* of the blackcap was wrongly termed by Bechstein a call-note: it is only so much of a call-note as an alarm may be.

In all the above instances (except perhaps in that of the bluethroat) the rattling alarm or the *chick* or *tack* are alarm-cries, and they blend in small degrees one from the other. The common spotted flycatcher utters after its young are hatched, and not often before, a short alarm, *chick*, which then often succeeds the ordinary call-squeak of the bird. This *chick* is practically the same note as that of the redstart, and the preceding squeak is precisely like the call-squeak of the young robin and the call-note of its own young. Sterland wrote that the whinchat utters as a call-note *chat chat chat* (*Birds of Sherwood Forest*, p. 69). This cry, which I know well, is very like the *tack* of the blackcap, and is equally an alarm. The same author (*op. cit.* p. 71) gives the call-note of the wheatear as a *chat:* a call or an alarm of the stonechat, as I have often observed, is very similar to that of the wheatear.

In the call-notes of birds of the thrush family are many instances of family resemblance. One of the most familiar of these notes is the long-drawn plaintive cry uttered by the migrating redwing. This note, which Sterland seems to have observed, is most noticeable at night in October and November, and may be imitated by whispering as loudly as possible *see you* as one word, through which the sound of the "s" is continued. The blackbird, in winter and when migrating, occasionally utters a cry which can only be distinguished by the closest attention from the cry of the redwing. I cannot think that this cry (which is a single note never repeated in quick succession) is, as Mr. Harting suggests, like the call-note of the fieldfare, which is rarely uttered except as a succession of short cries (*Birds of Middlesex*, p. 33). In the common song-thrush and in the American robin this note seems to be abbreviated to a shorter and somewhat harsh chirp; this, however, in both species, is sometimes prolonged to a much greater resemblance to the redwing's cry. In July and August young blackbirds, when disturbed, often utter, as they fly away, a short squeak or chirp somewhat plaintive in tone, and exactly resembling, except in its briefer length, the cry used by the bird when migrating in late autumn. The young of the American robin, which, from their colours and manners, would in England be assuredly mistaken for young blackbirds, utter, at the same season and age, precisely similar sounds upon the like occasions. I have never known a blackbird, when flying high and steadily to the south-eastward (as do the redwings) over Gloucestershire in autumn, and apparently migrating, utter its alarm-note, such as Herr Gätke records as uttered by the blackbird when passing at night the lighthouse at Heligoland. Probably near the lighthouse a passing bird would utter its alarm. The note I have heard from birds migrating has been a cry closely like, but distinct from, that of the migrating redwing. I have also heard this variation of the cry at night. The blackbird sometimes utters it by day in the same season, when entering a thicket. This note of the redwing is described by Dr. A. G. Butler as *phee-ur*, the call-note of the North American bluebird, a distant connection of our thrushes; " which note," he writes, " I have also heard uttered exactly in the same way by the redwing" (*in litt.*). Curiously enough, the cry employed by the tree-pipit when migrating is not unlike that of the redwing; in fact, it has just the same tone, but is shorter and less loud. The same cry in the meadow-pipit is yet more brief.

The blackbird, when suspicious of danger, utters at intervals of a few seconds a short full note, sounding like the word *quilp*. This is especially the case in autumn and winter. The note, as I have stated, is often raised in pitch, and the rapidity and continuance of its repetition are often increased until the common rattling alarm of the species is produced. But the redwing thrush, when suspicious of danger (at least in winter), utters the *quilp* as an alarm in precisely the same manner and tone as the blackbird. This I have many times observed; but I have never heard the redwing make any attempt to repeat the sound rapidly, or in varied pitch, in the manner of its near relation, the blackbird. The American robin has the same note of suspicion. The cries of the nestling young of both blackbird and thrush are much alike; and the young of the former species, when able to fly, begin to acquire a new character, namely, the first indication of a rattling cry. The cries of the nestlings of these species are more like that of the adult thrush than that of any other British bird.

The young nightingale, able to fly, utters a cry exactly like, but less loud than the cry of the young blackbird of the same age. When the male blackbird has lost his mate, or his nest has been destroyed, he utters at intervals a long and plaintive cry, of very high pitch, which, like the shorter cry of the migrating redwing, descends in about the interval of one half-tone. The cry might be whistled thus:

This note is also uttered by the male blackbird before rain in late autumn, and during prolonged frost in winter. On each of these occasions the robin sometimes utters a note which is the exact counterpart of that of the blackbird, but is much less loud, and is apparently higher in pitch. Dr. Butler writes: "A female robin about to leave her nest, in my garden, uttered the high distress-note, to which the male replied, and simultaneously with her leaving the nest he entered it, and brooded on the eggs" (*in litt.*). The American robin uttered a similar but shorter sound when disturbed by me near its young. Dr. Butler's captured redwings had the long, high distress-note of the blackbird (*litt.*). Layard wrote that the Cape chat-thrush (*Cossypha Caffra*), the Cape robin, sings like ours, and has similar manners (*op. cit.* p. 224). The common call-note of the robin, often heard before the song on still days in autumn, and also frequently used when the young are about, is a short squeak, less loud than the call of the male and female blackbird to each other, but of much the same tone. In autumn the young blackbirds of the year often utter this cry, which is also somewhat like the call of the common fly-catcher, and almost equally resembles, but is louder than the common cry of the young redstart. In fly-catcher and robin, at least, the sound is accompanied by a flirt of the tail. The nightingale utters to its young a squeak closely like the call-squeak of the robin, and I have heard it employed in conjunction with the alarm-croak of the bird, in the same circumstances. Bechstein observed that the European greater nightingale (*Philomela major*) has call-notes like those of the common nightingale

(*P. luscinia*), as well as a similar song (*op. cit.* p. 221). The same authority states that the black redstart's call-note, *fitza*, being very similar to that of the common nightingale, has given rise to its name of "wall-nightingale."

The common redstart not only has the alarm-cry, *chick*, like that of the common fly-catcher, but this note, as I have before stated, is usually preceded by a little whistle in the interval of about a fifth—

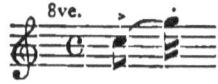

slurred upward, and sounding like the word *tewy* or *cooi*. This little whistle was common to all the male redstarts I have seen for some minutes in June. It is quite different from all the notes just mentioned, but is closely like the call-notes and danger-cries of the young and adult chiffchaff, and those of the full-grown and the adult willow-warbler. The white-throat utters the same cry, but at a much lower pitch, as a danger-signal. The nightingale (so far as I have observed in very many individuals) frequently commences a phrase with this cry, but in a tone much less loud than the succeeding notes; in fact, so softly that sometimes an observer must be not more than a chain or two distant in order to detect it; and I have heard it frequently uttered by nightingales attending to the wants of their young. Other birds, especially mimics, utter the cry in their songs; but the fact that in the nightingale it is addressed to the young, and in the song has precedence of other notes, indicates that it is the more natural to this species. The yellow warbler of North America (*Dendroica æstiva*) exhibits affinity to the willow-warbler of Britain, not only in size, form, colour, and habit, but also in voice. The song suggests the idea that a willow-warbler is commencing his song, but never attaining more than the first four or five notes. The song is sometimes prolonged by one or two notes, and then the resemblance to the willow-warbler is even more apparent. I have observed many yellow birds, both in Montreal and in Vancouver. The female is exactly like the willow-warbler, but has a faint light streak over the eye. The call-note of this bird, young or adult, resembles that of the young willow-warbler.

The fact that alarms addressed by birds to their young are practically invariable in character, though varied in vehemence, leads one to conclude that *tewy* is employed both as a call and as an alarm by the nightingale. The alarm-croak of the nightingale, so frequently uttered when the young are near, is produced in much the same but a softer tone by the sedge-warbler upon the same occasion. The lesser whitethroat has a similar low *currr*, used as an alarm.

SONGS OF BIRDS OF THE THRUSH FAMILY

The thrushes and their near allies are perhaps the sweetest singers. Their songs may be roughly divided into two classes, namely, songs in which set intervals of pitch and sweet tones predominate, and songs in which imitations of notes of other birds are the most noticeable. The former are generally the more melodious. I know of none of their songs which may be said to consist wholly of repetitions of call-notes. By set intervals of pitch I mean such definite intervals as may be readily perceived, but which are not necessarily found in the diatonic scale, although all birds—except perhaps the mute swan —have a vocal range of several tones. The music of bird-song will be discussed in a later chapter.

The mistle-thrush habitually repeats one musical strain many times in successive phrases before proceeding to another; and this habit is most often to be noticed in early spring. The song of the American robin has the same character; but whereas the repeated musical phrases of the former bird generally consist of from three to five notes, those repeated by the latter are generally of only two, or sometimes three notes. The following transcripts of the notes sung by two "robins" (*i.e.* thrushes) near my rooms at Vancouver will convey a good idea of the monotonous character of the performances of this species. The strains are marked by double bars; and each was repeated from six to a dozen times before the bird attempted a variation. (The pitch is not material.)

Though the British blackbird never sings in this barbarous fashion, it may be sometimes heard to repeat the same strain several times in succession; and this most often occurs in birds still wearing the infantile flight-feathers, which are shed at the second moult. (See also page 144, *post*.)

The song of the ring-ousel (which, however, I have only heard on one occasion) appears to be intermediate between those of the mistle-thrush and the blackbird, the phrases being less extended than those of the latter, and of wider range, and containing more slurs than the song of the former. Both blackbird and mistle-thrush habitually break from their full notes into some high shrill squeaks, which end their phrases. The nightingale follows the same method, but only in a slight degree. Bechstein has suggested some of its phrases by these words:

Tiŏ, tiŏ, tio, tio, tio, tio, tio, tix.
Tzu tzu tzu tzu tzu tzu tzu tzu tzu tzi.
Dzorre, dzorre, dzorre, dzorre, hi.

In each of these, and other similar examples stated by this author, the final syllable well suggests the frequent short concluding note of the bird. There is not a close resemblance between the songs of the blackbird and robin; yet each is a succession of full tones, and in this respect is like the full song of the blackcap and some of the strains of the nightingale. Mr. Herbert C. Playne, who has heard numbers of garden warblers near Oxford, informs me that the song of this bird is very like that of the blackcap.[1] The long notes of the nightingale, by means of which most people dis-

[1] See also Harting, *Birds of Middlesex*, p. 50.

tinguish its song, are represented in another bird, the wood-warbler, which, in the breeding season, cries a repeated full whistle, *kew, kew, kew,* seemingly as a signal of danger.

The mistle-thrush is an imitative singer, but its imitations are never uttered with a full voice: they are only heard after the full notes have been concluded, and they are uttered in a much softer tone, so that a listener cannot distinguish them unless he occupies a position fairly near the singer. They are often very good imitations. The blackbird also is slightly imitative; and in him, again, the imitations are most often (though not always) uttered after the full tones have ceased, when they form a continuation of the phrase. It is in the brown thrushes that mimicry is most evident—in the American mockingbird and catbird, for instance, as Wilson has so well recorded. The common song-thrush is the most imitative of its tribe in the United Kingdom. It sometimes sings short phrases composed only of whistled notes repeated, sometimes several times successively, in the same intervals of pitch; but these short phrases are at once distinguishable from the much longer and more irregular phrases of the blackbird, which also are of a fuller tone. The first songs of the young thrush and the young blackbird

—heard faintly sung in September, or later—are much alike, and seem to be aimless repetitions of whistled tones and slurs. However, traces of mimicry are soon observable. Wilson observes (*op. cit.* vol. i. p. 234) that the notes of the American ferruginous thrush (*T. rufus*) "have considerable resemblance to those of the song-thrush of Britain." Mr. Hudson tells us that the whistle of the Argentine blackbird is sometimes mistaken by Englishmen for that of the similar home bird (*Idle Days in Patagonia*, p. 155). Gould observed that the Australian little shrike-thrush (*Colluricincla parvula*), which may be said to be one of a class intermediate between the thrushes and the shrikes, has "a fine thrush-like tone, very clear, loud, and melodious" (*Handbook to Birds of Australia*, p. 225). The same authority remarks (*op. cit.* p. 279) that the song and call-note of the Australian scarlet-breasted robin (*Petroica multicolor*, Gould) "much resemble those of the European robin, but are more feeble and uttered with a more inward tone." He also states (*ibid.* p. 281) that "the red-capped robin (*P. goodenovii*, Gould) has a peculiarly sweet and plaintive song, very much like that of the European robin, but weak and not so continuous." Dr. Saxby heard wheatears which were excellent mimics (*Birds of Shetland*, p. 68). Sweet had a caged whinchat which was equally proficient (Bechstein, *Natural History of Cage Birds*, p. 243). I have heard several imitative stonechats. Bechstein quotes Sweet as giving this bird the same character, and Yarrell has the like observation (4th ed. vol. i. p. 281). Mr. Harting remarks that the garden warbler sometimes commences its song like a blackbird (*Birds of Middlesex*, p. 50). The blackcap is imitative; but, like the blackbird, it sometimes postpones its efforts in this direction to the end of the phrase. The whitethroat and lesser whitethroat also are imitative. I have never heard an imitation uttered by either chiffchaff, willow-warbler, or wood-warbler.

THE SYLVIINÆ, OR WARBLERS

Judging by resemblances of voice, I should place the chiffchaff and willow-warbler with the whitethroat at the head of this list. I have described how the nightingale calls its young, and also commences its phrases, with the utterance of a little upwardly-slurred whistle, like the alarm and callnote of the chiffchaff and adult willow-warbler, and also like the first of the two alarm-cries of the redstart near its young. The whitethroat has an alarm which is lower in pitch and harsher, but is slurred upward in just the same way as that of the willow-warbler and chiffchaff. It may be faintly suggested by the word *cruu-ee* pronounced rapidly and in this interval:

The lesser whitethroat resembles the nightingale (and sedge-warbler alarmed for its young) in the frequent utterance of a low alarm-croak, something like the word *currr*. It also resembles the nightingale in one feature of its song—the production of a long, full, bubbling note, closely like, but fuller than the long rattling song of the cirl-bunting. This song-note I have heard uttered by lesser whitethroats between Oxford and Witney, Aust Cliff and Gloucester, Andover and Southampton: I therefore presume that the exclamation is of general occurrence where the cirl-bunting also abounds. The nightingale repeats it more or less frequently. The redstart seems always to abbreviate it. Mr. Harting states that the Dartford warbler's song is not unlike that of the whitethroat (*Birds of Middlesex*, p. 54). The songs of the reed-warbler are of much the same character as those of the sedge-warbler. Mr.

Howard Saunders places the species in close alliance; and the late Dr. Bree wrote that the call-note of the marsh-warbler (*Sylvia palustris*) is similar to that of other reed-warblers (*Birds of Europe*, vol. ii. p. 73).

In regard to voice I should place the hedge-sparrow near the robin, because the call-squeak of the former is like that of the latter.

It might be well in this place to recapitulate in a concise way the cries connecting the various forms already mentioned. In the *Corvidæ* the croak of the raven is changed to a loud *corrr* in the crow, *caw* in the rook, *cah* in the jay, and a modified cry uttered as an alarm by the starling. The jackdaw utters it as a vehement alarm. The starling's notes of passionate song are high toneless squeals uttered at the end of the phrase, and often accompanied by a flapping of the wings. These notes separate this bird from the crows to which it is related by its alarm and the cries of its young. They form a link with the thrushes. The rattling screech of the mistle-thrush is uttered by its young, and is modified progressively in the common thrush, ring-ouzel, fieldfare, American robin, and blackbird, and seemingly in the South African olivaceous thrush. The robin's rattled cry has points of resemblance to that of the blackbird. The sound is not represented in the nightingale, but the single *chick* of the redstart and the *tack* of the blackcap may be modifications of the same utterance; and the latter bird especially, when much alarmed, repeats its note many times in succession.

The call-note of the migrating redwing is heard, slightly modified, in the blackbird, and greatly abbreviated in the common thrush and American robin. It also seems to have an analogue in the cries of the migrating tree- and meadow-pipits.

The cries for food of the young blackbird and young nightingale, when both have left the nest, are practically the same, differing only in force. The common short squeak used as a call by blackbirds, especially in autumn, is of much the same character as the call-squeaks of the robin and nightingale—cries which can hardly be separated from the calls of the spotted flycatcher and hedge-accentor. The young of the American robin have the same squeak as that of their close ally the blackbird, which has just been mentioned. The robin and blackbird have exactly the same cry or wail when the nest has been destroyed, or a mate lost, though the latter bird utters it much the louder. In the blackbird this note is repeated in severe weather in autumn and winter. The redwing has a single alarm-cry exactly like one of the blackbird, which is absent in fieldfare, thrush, and mistle-thrush, but occurs in the American robin.

The young redstart and young robin have at one period (just after leaving the nest) the same cries. In the redstart and nightingale we find a cry not uttered by blackbird, American robin, mistle-thrush, thrush, or robin, except as an imitation. It is the little whistle—

This is an alarm-cry of redstart, nightingale, chiffchaff, and willow-warbler, being especially prominent in the three last; and it is uttered in a modified tone and at a lower pitch as an alarm by the whitethroat. In the young willow-warbler the cry is usually heard as a whistle on the first of the two notes here indicated; but upon the excitement of fear the final note is added. The alarm-croak of the nightingale seems to be slightly modified in the lesser whitethroat and sedge-warbler.

There are considerable points of resemblance between the songs of several of the thrushes and the warblers, which I have discussed, and these resemblances appear to exist among certain foreign species. Amongst the former imitation reaches its fullest development in the American mocking-bird and the common thrush, and amongst the latter as represented in Britain, in the sedge-warbler (or in the marsh-warbler).

The wagtails, larks, and pipits seem to me to be more closely allied by voice to the thrushes and warblers than are the titmice, which Mr. Howard Saunders places next to them. Yarrell remarked truly that the call of Ray's wagtail "is more shrill than that of the other wagtails" (*op. cit.* vol. i. p. 382). The note of the grey wagtail, as he observed, differs from that of the others. The call or alarm-squeak of Ray's wagtail is like that of the tree-pipit, but is more shrill. The calls of the flying young tree-pipit, meadow-pipit, and pied wagtail are so much alike as to be almost indistinguishable. The pipits seem to be allied to the wagtails partly by flight, call-notes, and the length of their hind-toes. Yarrell observed that Richard's pipit has habits like the others, waves its tail like a wagtail, and has a long hind-claw like a lark (*op. cit.* 1st ed. vol. i. p. 399). The common tree-pipit and meadow-pipit have somewhat similar voices. Yarrell wrote of the rock-pipit, that "in its habits, mode of flight, and song, it so closely resembles the tree-pipit and the meadow-pipit as for a long time to have been confounded with them" (*op. cit.* vol. i. p. 394). The water-pipit (*Anthus aquaticus*) is described by Bree as singing like the tree-pipit (*op. cit.* vol. ii. p. 167). The songs of the tree-pipit and meadow-pipit, though delivered in much the same manner, have few tones common to both : the former bird has much the greater variation, the most frequent phrase being as follows :—*Chee chee chee chee cechaw eechaw whee whee whee whee whee whee ;* or *eechaw eechaw chee chee chee chee judge judge judge judge whee whee whee whee,* and so on. The meadow-pipit, on the contrary, rises crying, *chûwick chûwick chûwick*, repeated many times, and descends singing, *tsee tsee tsee* repeated ; or else it changes the accent from the first to the second syllable in the first cries, and ascends with *chuwíck chuwíck* repeated, with the same ending as before. Bree wrote that the cry of the tawny pipit (*Anthus rufescens*) "is very like that of the short-toed lark" (*op. cit.* vol. ii. p. 177). Mr. Saunders places the larks between the crows and the swifts ; but they seem more closely allied to the pipits, both by structure and habit. The British skylark resembles the tree-pipit not only in the habit of flying when singing, but also in the final notes of its early spring songs; these notes consist of a somewhat plaintive, prolonged, and repeated whistle, descending in pitch during its utterance. This note is abandoned by both skylark and tree-pipit towards the end of the season of song, and the habit of flight in song is at the same period less frequently exhibited. Though this whistle is not the common call-note of either adult skylark or pipit, it is similar to the call-note employed by the young skylark, not only during the period of helplessness, but on through the first winter. It is a slurred whistle which may be written thus:

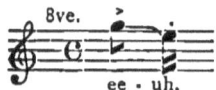

The common call-note of the adult skylark may be written *worry you* pronounced very quickly. Wilson, writing of the American shore-lark, states that in autumn "they fly high in loose scattered flocks, and then have a singular cry, almost exactly like the skylark of Great Britain" (*op. cit.* vol. i. p. 85). They are said to mount high and sing on the wing. Yarrell has the following:—" The song of the cock is lively and not very loud, and is more generally delivered when the bird is standing on some elevation than when on the wing, though at times an observer might fancy he was watching the characteristic flight, if not listening to the notes, of our own favourite at home" (*op. cit.* vol. iv. p. 509). The Abyssinian larks were said by Bruce to sing like those of Europe (*Dom. Hab. Birds*, p. 300). The South African lark (*Mirafra nævia*) sings in the air as well as on a tree (Layard, *op. cit.* p. 525). The Latakoo lark (*M. cheniana*) "somewhat resembles in its habits the English skylark, rising in the air early in the morning, with the same fluttering flight, and singing all the time" (*ibid.* p. 529). Thus we find American, European, and African larks agreeing in their manner of singing.

The buntings (*Emberizidæ*) appear to be allied to the pipits by their notes, in some species by the possession of a rather long hind-claw, and other features. The call-note of the young yellow bunting, and that of the young and the adult cirl-bunting, are not unlike the call-squeak of the tree-pipit. Yarrell quotes the late Mr. Proctor of Durham in relation to the snow-bunting as follows :—" I have often seen him, when coming from the nest, rise up in the air and sing sweetly, with wings and tail spread, like a tree-pipit" (*op. cit.* 1st ed. vol. i. p. 432). In the cirl-bunting the call-note of the young is uttered by adults in the breeding season ; and this bird somewhat resembles in its plumage the shore-lark. The buntings are, at any rate, much more allied by voice to the pipits than are the finches, which are often placed nearer to them. The song of the cirl-bunting does not vary in pitch during utterance. In this respect that of the yellow bunting is sometimes the same, but in Gloucestershire, and in the adjoining counties, it generally rises in pitch as the phrase proceeds; and as a rule this is followed by one, or two, shriller notes. An idea of the song has been suggested by the words, *A little little bit of bread and no cheese.* Bechstein thought the song of the corn-bunting shorter and less soft than that of the yellow bunting (*op. cit.* p. 117). The song of this bird resembles the noise made by shaking a few small keys together in the hand, beginning slowly and ending rapidly. The same writer thought the warbling of the ortolan had some resemblance to that of the yellow bunting, but with the last notes much deeper. Dr. A. G. Butler writes: "The song of the ortolan bunting is a mere repetition (about five times) of one note, in a bunting-like song, brief and unmodulated" (*in litt.*). Bechstein observed that the song of the European *Emberiza cia.* (Linn.) is shorter and clearer than that of the English yellow bunting (*op. cit.* p. 118).

Thus we find that several continental buntings sing like those found in Britain. The same resemblance may be observed in America. Wilson described the notes of the Towhe bunting (*E. erythropthalma*) as something resembling those of the yellow-hammer of Britain, but mellower and more varied (*op. cit.* vol. i. p. 185). The song of the American black-headed bunting is described as consisting of two notes, the first repeated twice and the second three times, *chip chip, che che che.* In form and manner these birds very much resemble the yellow-hammer of England (*op. cit.* vol. i. p. 54). I have often heard reed-buntings sing songs of this character. As a rule the songs opened with three notes, *cheeo, cheeo, cheeo,* followed by a short rattling repetition of a single cry, exactly like that in the song of the yellow bunting. Variations sometimes occurred in the termination. Mr. Harting states that the call-note of the snow-bunting is something like that of the linnet (*Birds of Middlesex*, p. 75). In *The Zoologist* for 1874, p. 4177, is a reference to the snow-bunting's tinkling note, from which it appears that this is used partly as an alarm and defiance.

Bechstein states that the agreeable song of the African indigo bird (*E. cyanea*) very much resembles that of the linnet (*op. cit.* p. 126). Wilson remarks that the rice-bunting (*E. oryzivora*), when migrating in autumn, utters a single *chink* as it flies (*Am. Orn.* vol. i. p. 200). The British reed-bunting, I have observed, sometimes utters, as at least a part of its song, *fink*, *fink*, but these notes may be the consequence of imitation, and not of inheritance. These notes, *chink* and *fink*, remind one of the *fink* of the chaffinch—a cry closely like the *fink* of the reed-bunting. But there is another note which has resemblance to the cry of a finch, and that is the common alarm-note of the adult yellow bunting. It is a short twitter, generally a double note, something like the words *did it*, pronounced as one, and which cry is sometimes repeated many times when danger is imminent. Let any one listen to the alarm-cries uttered by a flock of greenfinches and yellow-hammers, and he will at once perceive the resemblance between their alarms : the birds are not found together except during winter, when they occasionally seek food in the same spots. It may be said that the great titmouse cries *fink* exactly like a chaffinch, and that therefore, according to the line of reasoning adopted in this chapter, these two species should be placed somewhat near each other in classification ; but we must remember that the great tit has a variety of cries which are often varied in different individuals, whereas the greenfinch has but two distinct callnotes, *did it* and *tell tell*, and both sexes of the chaffinch have not more—the autumnal call, and the *chissick*, which the young chaffinches employ towards their parents, and which the sexes employ towards each other in spring.

Before tracing family resemblances in the voices of the great family of the finches, it would be well to retrace the scale of similarities from the pipits to the present point. The calls of the wagtails generally are alike, but Ray's wagtail has a call-note like the tree-pipit. Richard's pipit also resembles the wagtails in the waving of its tail. The pipits are all fairly alike in the manner of delivering their songs, and the transition from them to the larks is a very slight one. The final song-note of the tree-pipit in spring is closely like the final note of the skylark, and which, in the last species, resembles the call-note of the young. There are larks in America, Europe, and Africa which utter similar kinds of songs. The young cirl-bunting has a call like a pipit. The snow-bunting sings like a pipit, and the plumage of this species is somewhat like that of the shore-lark. Several continental buntings sing like those of Britain. The rice-bunting of America utters a *chink* as it flies; the snow-bunting has a tinkling note; and our own reed-bunting utters in its song a cry exactly like the *fink* of the chaffinch. The common yellow bunting utters as an alarm-note a double cry closely like that of the greenfinch.

NOTES OF THE FINCHES

The common call-note of the greenfinch, which may be rendered *didit*, or *tititit*, etc., is very similar to those of the goldfinch, siskin, and lesser redpoll, which utter the cry with degrees of repetition variable in the different species. Even the domestic canary, a true finch, notwithstanding the effects not only of artificial selection, but even of the artificial education of its voice, still utters as a call-note a twitter (the tone of which is suggested by the words *pretty dick*) of very much the same character as those of the greenfinch, goldfinch, siskin, and part of the call of the brown linnet. This appears to be absent from the voices of the house-sparrow and chaffinch. When the greenfinch, of either sex, is alarmed by a strange beast, or any bird of prey, in the neighbourhood of its nest, it utters a very distinct cry, an upwardly slurred whistle, of the same character as the *tewy* of the willow-warbler, nightingale, and redstart. This cry, in exactly the same tone, is commonly employed by the lesser redpoll. It is curious that upon certain occasions the common caged canary utters a note similar to this one of the greenfinch. It is often uttered in the presence of a stranger. The note of the greenfinch is sometimes uttered alone, but never in the song. The loud *zshweeo* of the greenfinch has already been described (*ante*, p. 36). Dr. A. G. Butler informs me that the *zshweeo* of the greenfinch is sung also by the brambling. The goldfinch utters a similar but very much shorter sound when attacking another; so also does the lesser redpoll, and in March the latter makes frequent use of the note. This sound has been alluded to in relation to the influence of combat (*ante*, p. 36). When the female house-sparrow is threatening a male who is "playing" to her, she produces a low rough note, of even pitch, which is seemingly produced partly by a very rapid vibration together of the mandibles. These birds, and especially the greenfinch, when taken in the hand, produce a similar sound. The penultimate syllable (to borrow a term) of the song of the chaffinch seems to be an abbreviation of the *zshweeo* of the greenfinch. It is early acquired by the chaffinch when recovering or developing the spring song.

In some country districts the song of the

chaffinch is said to express the prophecy : *In another month will come a wheatear.* Dr. Butler informed me that in Kent the "*wheatear*" was absent from the song ; and this I find to be very generally the case. In Gloucestershire the note is harder in tone. In Kent it sounds like " tissi-ear."

The short, full, contented-sounding note *jink*, which the yellow-hammer often utters from an elevated perch, may also be an abbreviation of this sound. The sounds made by the rattling of the bill during combat are much the same in the canary and in the house-sparrow. But to return to the greenfinch.

The greenfinch has one note which is frequently employed by the house-sparrow. This is a cry sounding like the word *yell*, often repeated. It may be whistled thus :

the small notes being hardly sounded at all. It is repeated as a call-note by young greenfinches, at least for some time after they can fly well ; it is employed with vehemence by the female greenfinch while her mate is feeding her near the nest ; and she then behaves in the manner of a young one on the like occasion; and it is a call-note used by the adults to express greater emotion than the ordinary call-note of *diditit*. This is proved by the readiness with which a flock obeys this cry in preference to the other, which is, however, generally sufficient as a call. I have proved even my own imitations of it (in a strange place) to be more attractive than the other cry, though this is uttered by a flock; but the experiment was conducted in winter, when the birds were much pressed for food. The lesser redpoll repeats the cry, *yell*, almost exactly like the greenfinch, and its young also utter it in the same manner as those of that species. The young of the brown linnet vociferate this cry when they are flying in flocks with their parents about July; and the young redpoll employs it, like the young greenfinch, when being fed off the nest. The same note is heard in the songs of the brown linnet, siskin, and goldfinch. By the house-sparrow it is employed in attempts at song; also as an alarm, and in the latter capacity with various degrees of vehemence; but in this species it resembles *tell* or *chell*, rather than *yell*, and is sometimes rapidly repeated.

The nestling young of the house-sparrow, brown linnet, and greenfinch have similar cries, quite distinct from the *yell* of the linnet, greenfinch, and redpoll, when able to fly. The call of the young house-sparrow is simply *chissick* or *chirri*, pronounced very briefly—practically the same as that with which the adult male calls the female, especially in early spring, when the cry is often repeated by him scores of times without variation. I have not heard the female utter the exclamation. The young sparrow makes great use of this note shortly after leaving the nest, at which period of life the young of all birds are most clamorous. The young chaffinch at the same age utters as a call for food a cry very like that of the young house-sparrow, but with a degree of resemblance varying in different individuals: in some it is so like this cry that it has deceived me, and I have imagined the exclaimer to belong to the latter species, before seeing the bird at a short distance. This cry, in a somewhat abbreviated character, is uttered by the adult chaffinches as a love-call. In the fourth edition of Yarrell's *British Birds* (vol. ii. p. 89) is a note recording that the house-sparrow was formerly called "Philip" sparrow from its note: this cry somewhat resembles *philip*, but is better rendered as *chissick*. In Montreal I observed that the house-sparrow, which is there very abundant, had all the cries of its British allies, and uttered them upon the like occasions. Bechstein states that the song of the tree-sparrow (this bird has a close resemblance to the common species) "is less monotonous than that of the house-sparrow" (*Nat. Hist. Cage Birds*, p. 139); Macgillivray observed that its note "is similar to that of the house-sparrow, but shriller" (*Brit. Birds*, vol. i. p. 352); and the like observation is to be found in "Yarrell" (4th ed. vol. ii. p. 85). "Its ordinary call-note is similar to that of the house-sparrow, but shriller" (Sterland, *Birds of Sherwood Forest*, p. 104). The rock-sparrow exhibits a tone of voice similar to that of the house-sparrow (Bree, *Birds of Europe*, vol. iii. p. 123). In South Africa we should find the same family resemblances, for Layard records, on the authority of Mr. Ayres, that the southern grey-headed sparrow (*Passer diffusus*) utters a note which "resembles the *chissick* of the English sparrow" (*Birds of South Africa*, p. 480). Bechstein has also recorded of the banded finch (*Loxia fasciata*, Linn.) or Indian sparrow— found in Africa—that its cry is similar to that of the common sparrow, and its song is not very different (*Nat. Hist. Cage Birds*, p. 3). The chaffinch is related to the house-sparrow by the call-note of its young (above mentioned); and this is an important feature, since the distinctive cries of young birds are generally invariable, and of universal occurrence in the respective species which utter them. The chaffinch, however, has another and a much more prominent note, which, although not heard in the young (and in this it resembles the *tell tell* of the house-sparrow), seems to be never absent from the adult; this is the *pink* (1st ed. Yarrell, *Brit. Birds*, vol. i. p. 462), *twink* (Knapp, *Journal of a Naturalist*, p. 272), or *fink*, which I have mentioned as an alarm and defiance, uttered by male and female (*ante*, p. 43). Dr. A. G. Butler suggests *chinck* as the most descriptive name for the note. The cry is not peculiar to the chaffinch: it is heard in the voice of the goldfinch and in the song of the brown linnet. Bechstein has recorded it as occurring in the song of the goldfinch. He writes of the twite (*Linota flavirostris*, Linn.), "The whole of the song somewhat resembling the first exercises of the chaffinch" (*Nat. Hist. Cage Birds*, p. 137). Macgillivray also writes that "its note is a *tweet* very similar to that of the chaffinch" (*British Birds*, vol. i. p. 337). Dr. A. G. Butler informs me that the song of the brambling resembles that of the chaffinch without the final "*wheatear*." There are many other instances of family resemblances between the notes of finches. Wilson has a note by Mr. Ord, that the call of the lesser redpoll exactly resembles that of the common yellow bird (*F. tristis*) of Pennsylvania (*American Ornithology*, vol. ii. p. 37). In *The Field*, number 2020, p. 430, Mr. F. H. H. Guillemard, in "Notes on the Ornithology of Cyprus," records that the cries of the serin finch found in Trooides Camp resemble those of the European bird, and that the habits of the two species are similar. Bechstein remarks that the song of the serin might be mistaken for that of the canary (*op. cit.* p. 155); and the same authority (*ibid.* p. 149) mentions that the goldfinch most easily imitates the canary, and also pairs with it—as is now well known. He also observes that the song of the Lapland finch (*F. Laponica*) "is very similar to the linnet's" (*op. cit.* p. 157); and he applies the same expression towards the song of the Angola finch (*F. Angolensis*). Wilson states that the note of the American pine-finch (*F. pinus*) "is almost exactly like that of the goldfinch" (*F. tristis*) (*American Ornithology*, vol. i. p. 276); and further, that the note of its near relation the purple finch is a single *chink* like that of the rice-bird (*ibid.* p. 119).

We will now refer briefly to the subject of the voices of the finches. The common call of the adult greenfinch is practically the same as those of the goldfinch, siskin, and lesser redpoll; and is heard slightly modified in the brown linnet and canary. This cry is not heard in the chaffinch or house-sparrow. The greenfinch has a rough song-note heard in a brief form in the song of the chaffinch. The goldfinch and lesser redpoll have a defiance-note of the same character, and a like incident is found in the house-sparrow. The call of the young greenfinch after leaving the nest is heard in the songs of the brown linnet, siskin, and goldfinch, and is a vehement call of the adult greenfinch. It is one of the chief warning-notes, and also a song-note of the house-sparrow. It is never uttered by the chaffinch. But the call of the young house-sparrow is reproduced, sometimes exactly, in the young chaffinch, and is always a soft love-call in that species. The prominent note of the chaffinch is *fink* or *tink*, which is also heard in the voice of the goldfinch and in the song of the brown linnet. Family resemblance is further traceable clearly in the voices of some European and American finches.

A comparison of the notes of raptorial birds generally reveals the same similarity of voice in birds physically allied. The titmice, woodpeckers, and cuckoos, whether of Europe or America, provide illustrations of the theory already stated. Pigeons and doves of Europe, America, and Australia respectively have the same habit of cooing. The young of the common pigeon, turtle-dove, and collared turtle-dove squeak in much the same tones. The cry of the chick of the common fowl greatly resembles those of the young of the partridge and pheasant. Birds of the same order, in Europe, America, and Australia, have sometimes similar cries and often similar manners. The young of the swan and duck utter whistling cries similar to those of the adult shelduck.

Enough evidence has now been adduced to show that frequently, if not in the great majority of genera, the physical resemblances existing between birds of a family have their analogues in similarities of voice in the same species. The instances mentioned in this connection are merely a small fraction of those available; yet so slight has been the attention paid by observers to this most interesting theme, that probably only a few of my readers will be able to criticise from personal knowledge the accuracy of my remarks on the voices of British species. Enough, however, has been said to demonstrate the family resemblances existing between the voices of many birds, though the species inhabit different hemispheres; and these resemblances suggest a common descent from general types.

The persistence of certain characteristic cries, over a wide area, was remarkably evidenced by the notes of two birds at Vancouver, B.C. One of these was the black-capped chickadee (*Parus atricapillus*), which is apparently identical with the marsh titmouse. On my using a "call," some four or five chickadees came around me, and, within the distance of two yards, poured out their notes, which were not distinguishable from the cry of the British bird, represented on p. 42. The second was a wren, apparently the "winter wren," in general appearance exactly like the European bird, and which sang practically the same song, and in the same manner, as its British prototype. It also made frequent use of a note identical with the coarse autumnal call-note of our wren. Thus, when nearly six thousand miles from Europe, half of which distance was occupied by the ocean, one heard what seemed to be the notes of two common British birds, though the birds of the two countries must have been, during countless generations, completely apart.

If the notes of these widely-separated tits and

wrens have been inherited from common ancestors respectively, the identity of tones just mentioned indicates the accuracy of the operation of heredity, or of filial mimicry; but, though we know that many analogous cries are often inherited, it does not follow that the persistence of particular tones is wholly due to the same influence. It is conceivable that the influence which first occasioned particular modulations has continually operated in their perpetuation. We hear comparatively so few of the noises which a bird hears in a feral life, that we cannot tell whether the sounds it produces resemble those to which it is most accustomed.

The family resemblances between the notes of allied birds must be due to heredity or to mimicry. It is evident that the notes of several species are perpetuated through many generations solely by inheritance; but the wide range of certain tones affords ground for the view that the persistence of characteristic cries in widely separate localities may be due to the influence of some equally persistent originals, towards which they have been modified by mimicry. Certain it is that even human ears can perceive resemblances between certain cries of birds, and other sounds frequent in the birds' habitats. This subject will be discussed in relation to imitation, and it is only mentioned here as suggesting a possible alternative to the marvellous force of heredity in the perpetuation of exact tones over vast areas.

This prevalence of tones has, I think, been demonstrated. It is rational to conclude that such family cries have been employed during a much longer period of time than songs, which are varied locally and individually; and that the original cries of the various kinds are recorded in their danger-cries and call-notes, and that the tones of later-developed cries, and modes of singing, are indicated in the first parts of songs, for these have the most generic characters.

The sounds next in order of prevalence differ specifically. These are songs.

Each distinct kind of birds may be said to have its class-tone, which we may consider as archetypal. In the crow family a simple *cah* or a croak seems to be the family cry. The gulls have a short loud scream, greatly varied. The thrushes, with the robin, nightingale, hedge-accentor, common titmice, wren, and many other insectivorous arboreal birds, utter a very short, and often very soft squeak for a call-note. The cry of the migrating redwing may be considered as of the same nature; and, in this connection, I should have stated on p. 101 that I have several times heard this note uttered in exactly the tone of the redwing by the mistle-thrush, apparently as a call between the sexes. These genera have different kinds of alarm-cries.

The raptorial birds have simple screams or short cries, sometimes consisting of very rapid repetitions of an extremely brief sound, which is probably the foundation-cry. In the duck family the young produce a *peet*, of the same nature as the *peet* of young doves, pigeons, pheasant, partridge, fowl, etc. —cries differing from those of most kinds of birds, but similar to the whistling notes of some of the *Anatidæ*, such as the shelduck when adult. Some simple form of whistled note may, therefore, have been the cry of the progenitors of these species; and in the same way the redwing's cry, and a simple chirp or scream, may have been respectively primary sounds in the thrushes and the falcons.

In the finches, the twitter of the greenfinch and its harsh song-note may be indicative of early modes of address. The coo of the stock-dove (*Columba œnas*), more simple even than the notes of the cuckoo, may be considered as a simple and early tone, whence have diverged the sometimes elaborate coos of the doves. The song of the ortolan, as described by Dr. A. G. Butler (p. 121), or that of the cirl-bunting (*ibid.*), may well be deemed the primitive form of a bunting-song. Likewise, the yell or pean of the peacock may be a survival of a cry of alarm, now adopted as a simple expression of defiance.

As already stated (p. 9), I do not, except in particular instances, refer to captive birds, and therefore I have not mentioned such interesting demonstrations of family resemblance of voice as one can easily perceive in zoological collections, and especially in that at Regent's Park, London. There one can hardly fail to be sensible of the strong family resemblance between the voices of allied birds, of the various races exhibited; and, in regard to those species which cannot be said to sing, this similarity is of scientific value, since in those species imitation seems to have less of that influence by which it so powerfully affects the notes of captive song-birds.

CHAPTER VIII

VARIATION IN BIRD-VOICES: ITS CAUSES AND EFFECTS

WE may safely lay it down as an axiom that vocal utterance is always subject to variation. Even in the newt and the snake such a variation occurs, though in the former it is an accidental result of altering degrees of violence in the contortion of the body, and in the latter it indicates varying intensity of fear or of hate. There is not much variation in the cries of the youngest birds; yet even here it occurs, and from causes analogous to those observable in the reptile, for when the suppliant has been satisfied he ceases to exclaim, and when very hungry he is most persistent in his cries. The same phenomena produce in him the same results, until he becomes sufficiently independent to abandon the use of cries for food, and then he simply calls occasionally for the companionship or aid of his parents.

There is a degree of variation due to physical development, such as one observes in the young of the mute swan and the duck, whose voices change when maturity approaches, in much the same way as human voices change, becoming lower in pitch and altered in tone. During all this period of development, from helpless infancy to capable maturity, variations of vehemence, and the extraordinary excitement of certain of the emotions, are frequently inducing proportionately pronounced modifications of the voice. Later on occur those variations which are always traversed before the first complete songs are produced. In some instances the latter are, as I have shown (Chapter V.), mere repetitions of a call-note, displaying variations in pitch; but in many species, and especially in those popularly considered to be the best singers, songs are developed from exclamations of much wider range; in fact, the songs of mature birds are not usually attained without varied exercises performed by the young in the course of developing their voices. While close attention is necessary to detect variations in the cries of nestling birds, it is also needed to observe alterations in the songs of many birds which are generally considered to repeat their phrases in the same respective characters.

The yellow bunting, chaffinch, willow-warbler, and chiffchaff, for instance, would by most people be considered to sing unvarying songs; but some evidence to the contrary will shortly be mentioned. The most general variations are in relation to the extent or rapidity with which an exclamation is uttered; less general methods occur in intervals of musical pitch; still less frequent is the acquirement of a note or tone quite distinct from that of the ordinary song of a species; yet even this last is a fairly wide-spread feature. Variation in the song of adult birds is not necessarily induced by erotism, for it occurs in birds which never see a mate; but there can be no doubt that the presence of a mate, or the desire for such presence at the breeding season, is the great stimulus which causes efforts towards extravagant song, either in the direction of a prolonged or a vehement utterance, or to the production of unusual cries. I have several times seen starlings utter their alarm-cries, apparently as a means of attracting the attention of the female bird; and sometimes execute an extraordinary devious flight—a succession of swoops and rushes, seemingly intended to excite the attention of other starlings, near which they had been singing. I have twice observed precisely the same behaviour, and in the like circumstances, displayed by the male greenfinch. In all these instances no hawk appeared, nor did the other birds near betray any sign of fear. Pigeons often swoop at each other in flight, as though in play : I am not sure that in these birds the action is induced by love ; but the same habit in the kestrel appears to be due to this emotion, when sometimes in early spring the birds call to each other, and swoop playfully at each other in the air. The cushat and the cuckoo also rise and fall in graceful swoops before others of the opposite sex, and some birds indulge in a jerky flight while singing : these have been already mentioned. The cock ruffles his quill feathers with his foot when playing up to the hen ; some woodpeckers rattle their bills in cracks in the bark of trees as a means of calling their mates. In these modes of expression, as in song, extravagance would attract the notice of a bird for which they were adopted ; and the fact of their wide-spread occurrence proves them to be thus attractive. Darwin mentions as "instrumental music" the rattling of quills by birds-of-paradise and peacocks, the scraping of wings against the ground by turkey cocks, the buzzing produced in a similar way by some kinds of grouse, and drumming by the North American grouse (*Tetrao umbellus*), which bird strikes his wings rapidly above his back in order to produce this sound. In view of the various movements of the wings, which so generally occur when birds approach their mates, it is questionable whether the "instrumental music" produced by these birds is not often an accidental consequence of an action intended as an overture. However, the domestic pigeon obviously strikes his wings together in flight as a call to his fellows. There is some evidence that bird-songs do not change in general characteristics except by a very slow process; but other facts suggesting the contrary conclusion will be subsequently related. Pliny wrote of the nightingale (*Nat. Hist.*, bk. x. cap. 43 : ed. Bostock and Riley): "Its note, how long and how well sustained! And then, too, it is the only bird the notes of which are modulated in accordance with the strict rules of musical science (?). At one moment, as it sustains its breath, it will prolong its note, and then at another will vary it with different inflections; then again, it will break into distinct chirrups, or pour forth an endless series of roulades. That there may remain no doubt that there is a certain degree of art in this performance, we may here remark that every bird has a number of notes peculiar to itself; for they do not all of them have the same, but each certain melodies of its own." Bechstein noticed differences in the songs of nightingales, as, indeed, would any close observer. He remarks: "For among these, as among other musicians, there are some great performers and many middling ones" (*op. cit.* p. 214); and quotes Daines Barrington for the statement that some are so very inferior as not to be worth keeping by the bird-catchers (*op. cit.* p. 219). A hundred years ago White of Selborne saw wood-warblers shivering with their wings a little, as they do now, when they made their sibilous, grasshopper-like noise in the tops of high beech-trees (*Nat. Hist. Selborne*, p. 36). It is evident from the number of instances of variation recorded by ornithologists that it occurs in the voices of all birds; nevertheless, the prevalence of characteristic phrases, betraying but slight trace of deviation in individuals, indicates that the tendency to vary is perpetually counteracted by others which conserve old characters in song. It appears that the emulation and vehemence of the males are the chief factors in the originating of variations, and that their emotions may in this respect chiefly be counteracted by the selection of female birds; for these, possessing limited voices, seem to have preferred males whose notes did not greatly differ from those generally prevalent in their respective species. The following are some instances of variation:—The hooded crow is said to vary its tone occasionally, producing two cries—the one hoarse, the other shrill (1st ed. Yarrell, vol. ii. p. 85). " In the breeding season it sometimes utters more musical sounds." Darwin remarks that the voice of the common rook is known to alter during the breeding season, and that it "is therefore in some way sexual" (*Descent of Man*, p. 375). It is, of course, employed to express sexual emotion, but is obviously of great use to the bird long before any such emotion is perceived.

Knapp observed that "birds of one species sing in general like each other, but with different degrees of execution"; and he recorded a diversity in the songs of the thrush (*Journ. Nat.*, p. 274). Bechstein states that no two species have a similar song (*op. cit.* p. 2); and it is correctly stated that individuals of one species differ widely in their accidental acquirements (*Dom. Hab. Birds*, p. 291). Rennie observed local variations in bird-song. Thus he remarked: "We have ourselves, in many instances, observed what might not inappropriately be called a different dialect among the same species of song-birds in different countries, and even in places a few miles distant from each other. This difference is more readily remarked in the chaffinch, dunnock, and yellow-hammer than in the more melodious species (*Edin. Mag.*, 1819). The chaffinches, for example, in Normandy were observed to vary from those of Scotland by several notes; and among the yellow-hammers of Ireland, England, and Holland we detected similar differences. We once heard a dunnock (*Accentor modularis*) in a garden at Blackheath sing so many additional notes to his common song, that we concluded it was of a different species, till we ascertained by watching the little musician that it was not otherwise distinguished from its less accomplished brethren." In January 1889, at Bournemouth, I heard two dunnocks, not far apart, which sang the same song—a phrase unlike that commonly uttered by dunnocks at Stroud, though of course easily recognised as the song of the same species. A dunnock at Paganhill, near Stroud, repeated, on 10th February 1892, the following very curious phrase (presto):

which I could not have recorded had not the bird persistently uttered it. On 15th April, at Frocester, a dunnock had chosen the following unusual strain for his song:

Of the chaffinch, Barrington wrote that those of Essex are the more esteemed by London bird-catchers. In Italy, as we learn from M. Montbeillard, the linnets of Abruzzo and of the March of Ancona are preferred. On this bird, see p. 127, *ante*.

What seemed an interesting example of progressive variation in the song of a blackbird occurred directly under my notice. In 1888 a pair of blackbirds reared their brood in a nest only a few feet distant from the window of the dining-room at my residence; and the young ones must have heard, for an hour or two every day, the notes of a piano, and sometimes those of other instruments, played in the room. Early in 1889 a blackbird in the garden (the only male, and a bird born in the previous year) sang little more than the following notes, which he would repeat dozens of times at short intervals:

The first note was strongly accentuated, and it was

rather flat in relation to the others, which are fairly correctly represented in this bar. As spring advanced the bird acquired a few other notes, always uttered after the others, but the same phrase was still his main theme. Next year it constituted about one-half of the songs of the old blackbird in our garden; and in 1891 it was often repeated, also in 1892. In 1893 I heard the same phrase sung three times in succession by an old blackbird in the garden. Blackbirds are with us throughout the year, and always roost in the same spots, which circumstances give occasion for the surmise that the same singer was with us throughout four years, and elaborated his notes from one kind of song to many others.

Bechstein also observed that the song of the chaffinch varied in different countries (*Cage Birds*, p. 35). Prof. Newton remarks (in a note on p. 550 of Yarrell's *British Birds*, 4th ed.) that to the best of his belief "the call-notes and songs of some other species—as the wheatear and the redstart—differ with the country in which they are heard." Local variations in the song of the yellow-hammer have been observed by Mr. W. Ward Fowler, who wrote: "The yellow-hammers in South Dorset in 1886 struck me as singing in a different manner from the Kingham birds, though it would be impossible to describe the difference. I think I have noticed the same in the case of the chaffinch" (*A Year with the Birds*, by an Oxford Tutor, second edition, p. 173).

A peculiar variation of the song of the starling has been already noticed (pp. 83, 84).

On 3rd June 1891, while walking from Chepstow to Tintern—some five miles distant—I listened to many chaffinches which were then in full song, and I stayed for a few minutes near six of them. These chaffinches sang a longer phrase than the usual run of those about Stroud, which is twenty miles from this locality,—the increase being readily detected in a double occurrence of the common penultimate and last syllables, which may be readily distinguished. In the chapter treating of music some variation in the notes of the chaffinch is recorded.

The varying degrees of excellence in nightingales has been noticed. Mr. Harting observed that the strain of the blackcap "is somewhat varied in every repetition" (*Birds of Middlesex*, p. 49). Quoting Mr. J. V. Stewart, he gives the song of the willow-warbler in Donegal as a succession of ten whistled notes descending from E to the C below its octave, in the key of C. This clearly differs from the song heard in Gloucestershire, which is rendered on p. 52, *ante;* but often the third, fourth, and fifth notes are considerably—sometimes a minor third —higher than the first. He gives the song of the yellow bunting as—

But in Gloucestershire the first notes ascend successively, so that the phrase follows nearly this course :

The complete phrase, however, is not usually heard until spring is advanced. Gilbert White noticed that some redstarts " have a few more notes than others " (*op. cit.* p. 55). I also have observed this, and that the phrase is lengthened during May, being more extended at the end of the month than at the beginning. The great tit, coal tit, and marsh tit often vary their songs by the utterance of some unusual note, which is, for some minutes, frequently pronounced. Bechstein noticed that the goldfinch varies in its frequency of uttering the cry *fink* in song (*Cage Birds*, p. 148). In the fourth edition of Yarrell it is recorded that among linnets there are great differences in individuals as singers (vol. ii. p. 155). Sterland wrote of the starling, that the woods "resound with their prolonged whistle, alternating with an oft-repeated gurgling note" (*op. cit.* p. 126); but I have noticed that this prolonged whistle is heard much less often in autumn than in spring. I once heard a robin, apparently in good health, utter a strange cry, *squee*, much prolonged, and somewhat resembling the low *currr* alarm of the lesser whitethroat, but totally unlike any note I had heard uttered by a robin. It was repeated several times at intervals, and seemed to be intended as a call-note, or as an alarm. The bird was about ten yards distant. In the following year, in autumn, I heard the same cry uttered twice by a robin in my garden, which was not more than half a mile from the scene of the former incident.

Mr. C. F. Archibald, of Rusland Hall, Ulverston, writes that in the spring of 1889 he heard a bird, which he was practically sure was a robin, sing an abnormal song like a tit's *tzee tzee che che che che che che che.* (The cry here alluded to is probably the long phrase of the blue tit.) The bird sang in the evening continuously, sometimes trying a few notes of the robin's song, but these were a failure (*in litt.*). I have often noticed that individual nightingales differ in the frequency with which they commence their phrases with the little upwardly-slurred whistle, which is one of their call-notes. Speaking very generally, but from much observation, I should say that in Gloucestershire the call is on an average uttered at the beginning of one in every four phrases of the nightingale. I have heard a willow-warbler (on 10th June) interrupt his usual mellow phrase with a sibilous repetition, like the song of the wood-warbler, but less prolonged.

Early in July, a male house-sparrow, which apparently lived near my bedroom, acquired a new note: it was seemingly a shout, and resembled the very rapid upwardly-slurred whistle, sounding like the word "twit," so frequently uttered in early spring by the male chaffinch (p. 43, *ante*). The sparrow repeated this cry every morning with the greatest persistence, and his manner of looking about during the performance indicated some pride in it. After ten days or so had elapsed he seemed to tire of the cry, and he gradually abandoned it. I never heard another house-sparrow utter this note. A neighbour also called my attention to the cry, which he thought to proceed from some uncommon bird. In June 1891 I heard at

Amberley, Glos., a male house-sparrow which had adopted a set order in the utterance of his call-notes. For an hour together he would cry at short intervals—*Tell, tell, chirri; tell, tell, chirri; tell, tell, chirri*, and so on. In May 1892 I heard at Corbett House, Stroud, a male house-sparrow crying very frequently *chirrirri* (or perhaps *chississick*), instead of the common call *chirri* or *chissick;* and he was addressing a female bird, as is usual when the cry here alluded to is employed. In 1890 Mr. S. S. Buckman informed me that during the two preceding spring seasons there had been near his residence at Oxlynch, Stonehouse, Glos., a cuckoo which could sing only *cuck*, instead of the usual cry of the species. Conversely, in *The Field* of 2nd July 1892, it is related that in 1891 a cuckoo, near the house of a correspondent, sang every day three notes instead of two; and the three were "perfectly distinct and of equal value." The three notes of this cuckoo and the triple-syllabled cry of the house-sparrow just mentioned may possibly indicate a tendency towards the construction of a phrase by the cuckoo and sparrow. The cuckoo begins to sing with the interval of a minor third, then proceeds to a major third, next to a fourth, then a fifth, after which his voice breaks without attaining a minor sixth. For an interesting quotation on this subject, see Mr. Harting's *Ornithology of Shakespeare* (1871), p. 150. The musical terms here employed are, of course, intended merely to suggest and not to define the intervals sung by the cuckoo; and although every one must admit their general suitability, great deviations from this regular transition are of common occurrence. The bird, early in May, very often sings a major second instead of a minor third; and sometimes at this period it utters any interval up to a major fifth. One bird, heard by me on 18th May, sang many times a major third, immediately followed by a major fifth. The song of the mute swan, uttered in the interval of a minor third, is given in Mr. Harting's *Ornithology of Shakespeare*, p. 202. I have heard the kingfisher repeat his squeaks in the same interval, apparently as a song; this was in early spring. Rennie made the following observation on the universality of variation : " Though birds of the same species very closely resemble each other in the general tenor of their song, individuals differ widely, both in the introduction to particular passages, the result probably of accidental acquirements, and in skill of execution, as well as in intonation." Wilson said he was so familiar with the notes of an individual wood-thrush that he could recognise him above his fellows the moment he entered the woods (*op. cit.* p. 291).

Variation in song is most noticeable at the end of the phrase, the first parts of phrases being more like common types than are the later parts. This is most observable in birds which do not imitate very much, nor greatly vary the character of their song; such are the nightingale, which so often preludes its song with one of its call-notes; the chaffinch, which particularly varies the two last syllables of its song; the greenfinch, which always commences a phrase with a rapid repetition of its *titit* call-note; the willow-warbler, which follows the same course of proceeding, and varies its sweet phrase towards the close. The reed-warbler seems to commence each phrase in much the same manner, and even the sedge-warbler often exhibits this method. I think we may safely say that the songs of all birds which can be said to sing become extended during the season of song, though the extension be but slight, as in the chaffinch, or the variation be merely an extension of an interval of pitch between certain notes, as occurs in the cuckoo. The phrases of the robin, starling, blackbird, thrush, brown wren, lark, redstart, yellow bunting, blackcap, lesser whitethroat,

whitethroat, greenfinch, all become prolonged as the season of song advances; though this feature, of course, is not restricted to the breeding season in all these species, for the adult robin and starling, at least, recommence their notes early in August, while the skylark, wren, and hedge-sparrow sing sometimes in autumn.

In treating of imitation, I shall have something to say about local variations and what caused them; I will, therefore, not now say more than that the local variation mentioned by Bechstein and by Mr. Ward Fowler, as occurring in the yellow-hammer and the chaffinch, may be clearly traced in some other species.

It should be remembered that there are variations in the rate of uttering notes, as well as in their actual pitch, a feature especially noticeable in the robin's *lit it it* alarm. There is often an extraordinary extent of diversity in the utterance of this alarm, and of the accent accorded to the notes. I once attempted to record this exclamation, as sung by a robin, and wrote large dots for loud notes, and small dots for soft ones, and left spaces to indicate the amount of time elapsed between them. Each line represents one utterance of the alarm, and should be read from left to right.

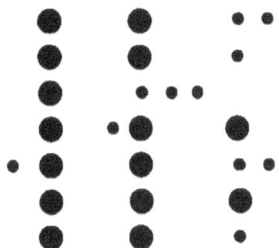

The phrase most often runs to a much greater length, thus:

● • • • • • • • • ● ● ●

CHAPTER IX

THE INFLUENCE OF IMITATION IN RELATION TO BIRD-SONG

IN the voices of animals, hereditary similarity, involuntary or passive imitation, and voluntary or active mimicry, blend insensibly from one to the other. The influence of resemblance is to be observed throughout the animate world. Mere physical resemblances to things or to organisms are of vital importance to vast numbers of creatures which they continually protect. Whether the primary causes of these resemblances were fear, and consequent efforts towards concealment, giving rise to ocular impressions, which have found physical expression through the agency of generation; or whether their origin may be discovered in the influence of the protection afforded by similarities of form and of colour, inducing selection by enemies, who would see first the most noticeable of their prey; or whether animals thus protected exercise an appreciative choice, or are instinctively impelled when seeking positions in which their imitativeness is most effectually protective—are themes foreign to the subject of this chapter. Nevertheless, we should remember that the distinction between appreciative choice and instinctive choice is rather narrow. Involuntary imitation, such as that of young birds learning the notes of their parents, is generally perpetuated by inherited tendency; and this, together with the protection afforded by imitation, no doubt chiefly perpetuates some of those common features by which each species of animal is distinguished. The power of filial mimicry is evidenced by the behaviour of the house-sparrow, which, though (as will be seen) endowed with imitative tendencies, has preserved its language on the American continent. At Montreal I heard the sparrows employing all the cries of the British birds (see p. 130).

Imitation is occasionally exhibited in the voices of animals other than birds. Several dogs have been taught to utter sounds which resembled words, and I have observed a remarkable instance of similar mimicry in a Dandie Dinmont terrier which belongs to my brother. When the terrier was about two years old, a friend stayed with us, who had with him a fine female mastiff. The little dog was somewhat awed by this great beast, which could easily have made a meal of him; but he was evidently very proud when allowed to accompany her for a ramble in the country. After the mastiff had been with us for a week or so, the terrier was heard trying to reproduce her baying, which was, of course, much lower in pitch than his bark. Several persons witnessed these attempts at mimicry, and all said how great efforts they appeared to be. Some days afterwards I heard the terrier essay the imitation, which he did very successfully. He raised his head and uttered a great bark, about an octave lower in pitch than his usual tone. All his breath was exhausted by the effort, and he immediately coughed, as though his larynx had been strained. I was informed that his previous attempts had also had the same effect. Weeks after the big dog had left, the little one showed signs of sadness—by lowering his ears, and in other ways—when her name was pronounced to him. It is difficult to say how far down the scale of creation we shall find the lowest occurrence of voluntary imitation. I used to keep fish in an aquarium, and I noticed that when one gaped any other near would be tolerably certain to gape soon afterwards, though none had gaped during the preceding half-hour. Fish, again, and some insects also, are particularly liable to follow each other; and here imitation may influence the choice of direction.

In treating of the influence of heredity, I have shown that the young of several species of birds inherit their cries; and indeed it is probable that, speaking generally, the cries of birds which have limited voices are inherited, and that those of what are commonly called " singing-birds " are perpetuated through the agency of mimicry. Certain it is that the call-notes of the fowl, pheasant, turkey, partridge, duck, goose, and common shelduck are inherited; those of the pigeon, cuckoo, crow and his allies, hawks and their allies, sea-birds, and others of limited voices, are probably inherited, but may be transmitted by imitation; while the cries of some of the finches and warblers are certainly perpetuated through the latter agency, and the songs of a large number of these last certainly are greatly affected by its influence. The effects of this influence are wide-spread amongst captive song-birds. Who has not heard a caged finch, or lark, blackbird, thrush, or starling which was said to have "caught," as it reproduced in its song, the notes of one or more neighbouring birds? Such instances are very common; and I do not believe that one of our native song-birds, reared in captivity and isolated from its species, will ever sing its natural song. Having, during the last nine years, paid great attention to the singing of birds, I can generally detect whether a loud singer, such as a thrush or blackbird, heard even at a distance, is a caged or a free bird. A canary, after once acquiring its song, seems less liable to imitate birds not in its immediate vicinity than is the linnet or the skylark; yet many canaries adopt strains from other birds, independently of the condition of the latter. At Vancouver many of the canaries had learned the invariable, short, but full-toned song of a certain bird abundant in all the vacant spaces in the city. Sometimes I could not tell which species was singing. It is to be regretted that many writers, in recording the imitativeness of birds, do not state whether they refer to caged or to wild individuals.

The crows, which are notoriously imitative in captivity, do not apparently reproduce when wild the notes of birds not of their respective species, although they possess the apparatus for singing, as we have seen (page 70), and although their powers extend to the reproduction of sweet whistles, as well as the tones of the human voice. The necessities of their lives in a feral condition may prevent them from exercising this power to its full extent. This idea is supported by the behaviour of the crows at Vancouver city, B.C., where these birds are accorded a judicious protection, and may be seen fearlessly promenading all the open spaces. Professor Hill-Tout, Principal of Buckland College, Vancouver, has kindly written for me the following note on these birds :—

"I would call these birds strong imitators of sounds in their wild state, and am inclined to think that they would equal the parrot in captivity if carefully taught. I have frequently heard them imitate the barking of a dog, and, less often, the sounds one hears in a fowl-yard. On one occasion, when in the country with a friend, I was surprised to hear, from a bush close by, sounds that strongly resembled human cries. We were for a time under the impression that the sounds came from an Indian camp, and we were much interested to find that they proceeded from a crow in a tree."

Bechstein mentioned incidentally that the nut-cracker (*Nucifraga caryocatactes*) can imitate the notes, but not the songs, of other birds (*op. cit.* p. 32). The jay is more imitative: at intervals it will introduce into its song "the bleating of a lamb, mewing of a cat, the note of a kite or buzzard, hooting of an owl, and even the neighing of a horse. These imitations are so exact, even in a wild state, that we have frequently been deceived" (Montagu, quoted by Macgillivray, *op. cit.* vol. i. p. 579). Yarrell and others also mention its imitativeness, with which, indeed, we are all familiar. Its near ally, the blue jay of America, is evidently a great mimic (Wilson, *op. cit.* vol. i. p. 27). None should need the citation of an authority for the statement that the starling is exceedingly imitative when wild or in a cage (but see Yarrell *op. cit.* 4th ed. vol. ii. pp. 229, 230). The jays are the most arboreal of the order to which they belong; and to their residence among trees may be attributed their frequent and elaborate use of the voice.

I consider that the mistle-thrush is a moderately good mimic, although its mimicry is uttered in a tone much softer than its ordinary full notes, and in consequence of this, and of the wariness of the bird, is difficult to hear. The thrush is a capital mimic, reproducing in its song a multitude of sounds borrowed from its avian neighbours. For an excellent description of the mimicry of the mocking-bird, see Wilson's work (vol. i. pp. 166, 167). The catbird is almost the rival of the mocking-bird (*ibid.* vol. i. p. 243). The story of a blackbird imitating the crowing of a cock has often been repeated, and probably owes its notoriety to the want of observation in writers who mentioned it; for this bird readily reproduces other cries, although they are usually performed in a faint voice. Rennie has quoted Kircher's assertion that the young nightingales which are hatched under other species never sing till they are instructed by other nightingales (*Musurgia*, cap. de Lusciniis: *Domestic Habits of Birds*, p. 277); and he has quoted the Rev. W. Herbert from White's *Selborne* (8vo, 1832) as follows:—" The nightingale is peculiarly apt, in its first year, when confined, to learn the song of any other bird that it hears. Its beautiful song is the result of long attention to the melody of other birds of its species"; and (p. 278) that the nightingale, when young, " will learn the notes of other birds, and retain them after it has heard its own species again." The Hon. Daines Barrington found the young robin equally imitative. " I educated a young robin under a very fine nightingale, which, however, began already to be out of song, and was perfectly mute in less than a fortnight. This robin always sang three parts in four nightingale. I hung this robin nearer to the nightingale than to any other bird; from which I conceived that the scholar would imitate the master which was at the least distance from him. I find it to be very uncertain what notes the nestling will most attend to, and often its song is a mixture. . . . I educated a nestling robin under a woodlark-linnet, and afterwards put him near a skylark-linnet, and the robin imitated the latter entirely" (*Domestic Habits of Birds*, pp. 275, 276). The redstart also mimics. Yarrell compared it to the blackbird as an imitator, and he recorded that in a cage it may be taught to whistle a tune (*op. cit.* 1st ed. vol. i. p. 238). The blackcap "is said to be an imitator of the notes of others" (*ibid.* 4th ed. vol. i. p. 420). I have frequently observed its mimicry. Gilbert White noticed that the sedge-warbler imitated other birds; and Yarrell observed that it did this in a somewhat confused and hurried manner (*op. cit.* vol. i. p. 377). Of the mimicry of this bird I have much to relate. The marsh-warbler (*Sylvia palustris*) "imitates with exactitude the notes of the goldfinch, chaffinch, blackbird, and many others" (Bree, *op. cit.* p. 74). Mr. Herbert C. Playne heard remarkable mimicry displayed by one of these birds near Oxford. Mr. Playne also informs me that the reed-warbler is an excellent mimic, he having heard many of them near

Oxford. I heard a bird of this species imitate fluently and accurately the cries of many birds living in its vicinity. Jesse (*Gleanings*, p. 172) observed that the whitethroat imitates the notes of the swallow and sparrow. I have often observed these imitations, and others less noticeable, performed by the whitethroat. The skylark is very imitative. Jesse whistled a tune to one which he had reared from the nest; and he heard the bird "inwardly whistle, or, in the language of bird-fanciers, 'record' it." Bechstein wrote that its young in cages readily imitate, but not generally the old ones (*op. cit.* p. 178), and that the Calandra lark can imitate all sounds adapted to its organs (*ibid.* p. 185). The common skylark is imitative, both when wild and when caged. A whinchat reared by Sweet from the nest learned songs of the whitethroat, nightingale, willow-warbler, and mistle-thrush, which it frequently heard singing in a garden near (Bechstein, *op. cit.* p. 243). Sweet wrote that the stonechat has a strong voice "to imitate the notes of another" (*ibid.* p. 244). I have heard the wild stonechat mimic very well. Yarrell recorded the imitativeness of these two species. The Rev. W. H. Herbert, above quoted, states that the young wheatear, whinchat, and others of the genus *Saxicola*, which have little natural variety of song, are no less ready than the nightingale, in confinement, to learn from other species (*Dom. Hab. of Birds*, p. 277). I have heard only one wild greenfinch imitate, and that one reproduced the *fink* and *twit* of the chaffinch; but in confinement the bird will repeat "the song of any fellow-captive" (Yarrell, *op. cit.* vol. ii. p. 106); and Bechstein said that it could be taught to repeat words (*op. cit.* p. 99). Rennie had a young male greenfinch which, in his opinion, from hearing the "call" of the sparrows out of doors, had acquired it perfectly; and, from hanging near a blackcap had also learned many of this bird's notes, though it executed them indifferently, perhaps from deficiency of voice. "Yet notwithstanding that he has thus learned part of the notes of three or four different birds, he can also utter the peculiar call-note of his own species, though we are pretty certain he has not heard it since he left his parents' nest, when only a few days old" (*op. cit.* p. 281). Barrington took a common sparrow from the nest, when it was fledged, and educated it under a linnet; the bird, however, by accident, heard a goldfinch also, and his song was therefore a mixture of those of the linnet and goldfinch (*op. cit.* p. 275). Mr. A. Holte Macpherson informs me of a sparrow in the possession of Canon ——— which was taken when quite young from a nest in a pump and brought up in a cage with canaries, and which sings just like a canary, only better (*in litt.*). Sterland reared a house-sparrow in a cage adjoining that of a skylark, whose song the sparrow learned, and its owner "often admired its surprising imitation." It would interrupt its song with the ordinary sparrow call-notes (*op. cit.* p. 115). Barrington educated nestling linnets under the three best singing larks—the skylark, woodlark, and titlark —"every one of which, instead of the linnet's song, adhered entirely to that of their respective instructors. When the note of the titlark-linnet was thoroughly fixed, I hung the bird for a quarter of a year in a room with two common linnets which were fully in song. The titlark-linnet, however, did not borrow any passage from the linnet's song, but adhered steadfastly to that of the titlark" (*Domestic Habits of Birds*, p. 273). One of his birds learned with equal ease the song of an African ally. "I therefore educated a young linnet under a Vengolina (*Linola Angolensis*, Brisson), which imitated its African master so exactly, without any mixture of the linnet's song, that it was impossible to distinguish the one from the other. This Vengolina linnet was absolutely perfect, without uttering a single note by which it could have been known to be a linnet. In some of my other experiments, however, the nestling linnet retained the call of its own species, or what the bird-catchers term the linnet's chuckle, from some resemblance to that word when pronounced. All of my nestling birds were three weeks old when taken from the nest, and by that time they frequently learn their own call-note from the parent birds, which consists only of a single note. To be certain, therefore, that a nestling will not have even the call of its own species, it should be taken from the nest when only a day or two old; because, though nestlings cannot see till the seventeenth day, yet they can hear from the instant they are hatched, and probably, from that circumstance, attend to sounds more than they do afterwards, especially as the call of the parents announces the arrival of their food" (*ibid.* p. 273).

It is evident that in these remarks Barrington alluded to singing-birds only, the newly-hatched young of some other orders of birds being able to see and run. He saw a linnet which had been taken when only two or three days old, and which he was assured had no call-note of any bird whatsoever; he heard it almost articulate the words *pretty boy*, and some other short sentences. He also heard a goldfinch, which had been taken at the same early age, and which had no call of its species, but sang only the song of the brown wren, which it had heard in the garden in front of the window where it was hung. I have heard a caged linnet warble to perfection the songs of the blackcap and brown wren. This bird was in a railway signal-box, in the midst of tall trees. Rennie stated that "the canary-bird, whose song, in its artificial state in Europe, is a compound of notes acquired from other birds, is able to learn the song of the nightingale, but is not able to execute it with the same power as the nightingale does" (*op. cit.* p. 278). Bechstein recorded that the siskin "imitates tolerably well the song of other birds, such as those of the tit, chaffinch, and lark" (*op. cit.* p. 153). Yarrell observed that individual bullfinches vary in their ability to learn. Even so poor a songster as the hawfinch will imitate when a captive. The Rev. H. A. Macpherson informs me they will pick up any sounds (*in litt.*). I hear from Mr. Lulham, a breeder and exhibitor of hybrid finches, that, as a rule, the songs of these birds are extremely poor: "they generally adopt a few notes from any birds they hear, but, with the exception of goldfinch and canary mules, their song is not very pleasing" (*in litt.*). Mr. W. A. P. Hughes informs me that "a young bird (finch) reared by hand, and not allowed to hear another bird, never learns a perfect song, but sings a series of disconnected notes, without any similitude to its parent's song. Young bullfinches or greenfinches, bred from the egg under canaries, learned the foster-parents' songs, and had none of the harsh notes of their actual parents; while young greenfinches taken from the nest when fledged, and then reared by hand, always had some of their respective parents' notes, although learning another song under a tutor. A goldfinch-canary mule with a pure goldfinch song, when two years old, learned the song of a linnet, and sang both songs alternately and mixed. The time when the young bird really picks up the song is when in the nest, and before it can feed itself. I have seen the featherless little birds singing; that is, I have seen their throats going, and heard their squeaky notes, as though they were practising" (*in litt.*). In support of these observations is the fact that the males of some of the finches (*e.g.* the greenfinch) feed the females near the nest during incubation, on which occasions the call-notes are abundantly employed, and hence attract the notice of the young, then exposed to the air. The Rev. H. A. Macpherson informs me that bullfinches try to pipe as soon as they can perch (*in litt.*).

The great grey shrike, whose variety of tones has

been mentioned, is said to imitate other birds for the purpose of attracting them within range of its attack (Yarrell, 4th ed. vol. i. p. 201). Yarrell noticed that the red-backed shrike has a note like the house-sparrow. This seems to be an alarm-cry resembling the word *tell*, and pronounced louder than the *tell* of the house-sparrow. I have only heard it used by the shrike in May. Bechstein recorded the imitativeness of both the woodchat shrike and the lesser shrike in confinement (*op. cit.*).

Mimicry is by no means confined to birds of the north temperate zone : even in Australia, where bird-song is not so fluent or melodious as in Europe, some excellent imitators are found. Dr. Stephenson stated that the *Menura Alberti* (Prince Albert's lyrebird), a bird about as large as a fowl, imitates with its powerful musical voice any bird which it may chance to hear near it; the note of the "laughing jackass" it imitates to perfection. Its own whistle is exceedingly beautiful and varied (Gould's *Handbook to the Birds of Australia*, vol. i. p. 308). Mr. A. A. Leycester had heard one which had taken up its quarters within 200 yards of a sawyer's premises, and had "made itself perfect with all the noises of the sawyer's homestead—the crowing of the cocks, the cackling of the hens, and the barking and howling of dogs, and even the painful screeching of the filing or sharpening of the saw" (*ibid. op. cit.* vol. i. p. 310). Gould described the superb lyre-bird (*M. superba*) as "sometimes pouring forth his natural notes, at others mocking those of other birds, and even the howling of the dingo" (*ibid. op. cit.* vol. i. p. 300). These descriptions of the menuras evidently refer to wild birds. The Tasmanian cow-shrike "possesses the power of imitating in an extraordinary degree: it may be taught to whistle tunes as well as to articulate words" (*ibid. op. cit.* vol. i. p. 178). Mr. R. Sterndale, of Whitcombe, Worcester Park, Surrey, has kindly forwarded to me an extract from a paper published in the *Journal of the Bombay Natural History Society*, vol. i. p. 28, 1886, by Mr. E. H. Aitken, "On the Mimicry shown by *Phyllornis Jerdoni*, the Green Bulbul," which bird he twice heard imitate many others. He says that on one occasion the bird, which he had alarmed, "began to abuse me in several languages." General D. Thomson, who spent many years with the army in India, writes that many birds in that country are very imitative, notably the crow and the minah (*in litt.*).

We have here records of the imitativeness of birds of many different kinds; and we find that in two species at least, goldfinch and linnet, every note is acquired by imitation, or rather, that none is inherited. I think we are justified in this general conclusion from the behaviour of only two or three individuals, for the cries of young birds are practically invariable specifically, nor do they vary individually except in one or two species; and we find also that the house-sparrow, greenfinch, and bullfinch learn all of their songs by imitation, and inherit none of them. How different in this respect are these birds from the rasorial birds, or from ducks, which, as we have seen, utter the same cries respectively, whether they have been reared artificially or naturally!

The receptivity of young singing-birds must be very great; but we should remember what a stimulus that faculty receives from the necessities of the young—their constantly recurring hunger, appeased by the parent, which often at the moment of feeding them utters some call or alarm-note; and their chilliness, removed by the arrival of the brooding mother. These influences are doubly powerful when the young are still blind, for then the feeble creatures can only by the sense of hearing perceive the approach of their helpers.

The fact that most of the imitative birds above

mentioned utter recognisable imitations when caged and secluded from their kind does not in any way prove that their imitativeness is solely caused by captivity or by modes of life incidental to that condition; although, when caged birds imitate, they generally do so without any encouragement in the form of desirable food from their owners. It does not indicate that the dispositions of birds are in any way altered by the change, except by the presence of abundant food, and a consequent enforced leisure. The bird is not thus changed, but its surroundings are altered, and its song is subjected to fresh influences, which naturally produce fresh results. The fact that in most species the songs of wild individuals are always of much the same characters respectively (*e.g.* brown linnet, greenfinch, goldfinch, bullfinch, blackcap, etc.), and that these birds, when confined in cages near others, abandon their own specific songs in favour of parts of songs of these neighbours, is strong evidence upon one point—the great value of specific notes to birds which normally employ them. It indicates the importance of every class-note in the voice of a bird, whether a call or a song, in the social economy of bird-life. The value of such cries must be one of the chief influences counteracting that imitative tendency which is latent, if not evident, in nearly every bird with any pretensions to song. Or it may be that the power of mimicry is exercised by wild birds generally in an unconscious endeavour to resemble others of their species—a similarity of voice being an important aid in the distinguishing of birds of each race, which more or less frequently seek each other. In announcing the approach of an enemy, singing for a mate, or calling to the young, a great variation from a common type of exclamation would certainly be injurious in effect, and would therefore be of rare occurrence. If a man be taken from his native country, and be placed with foreigners only, will he not learn and repeat their language? The bird does the same. We should not allow our regrettable ignorance of bird-song to lead us to conclude that because we understand hardly anything about it, the birds themselves can perceive no more meaning in it. Their fidelity to their songs when with their own species, even though dwelling within hearing and being sometimes the companions of other birds, and the sudden change of voice consequent upon a total change of surroundings, are facts not to be lightly passed over; they indicate that the language of the wild bird is to it as important as is the language of a wild man to him.

The detection of the imitations sung by birds in cages is rendered easy by the circumstance that the birds almost necessarily reproduce sounds heard around the cages, and hence familiar to persons in charge of the captives. Conversely, the imitativeness of wild birds would be difficult of detection by any one not familiar with the subjects imitated, whether these were cries of birds or sounds caused by other animals. I shall endeavour in subsequent pages to show how greatly imitative are many of our own wild birds, especially the sedge-warbler, starling, thrush, blackcap, robin, and skylark. Until extensive experiments have been made, we shall not be able to determine the extent to which imitation generally affects the call-notes of the young. We know that it is responsible for the character-cries of the goldfinch and brown linnet; and probably it is equally powerful in several species allied to them. However this may be, we are sure that songs are often completely ordered by this faculty, and if songs are thus influenced, call-notes also probably are similarly affected. Our present knowledge merely shows that the characters of the call-notes are more permanent than those of the songs, and that they are more influenced by physical inheritance or by filial mimicry. The call-notes, from being less variable than songs, both in individuals and in species, would seem to have existed for a much longer period, and to have thus been more liable to the influences of imitation. Had this always been so, we should find in the call-notes some traces of imitation; and the sounds most likely to have been imitated would have been those incidental to feeding or to obtaining food. This idea is partly supported by the behaviour of the common cock, who, when calling hens to corn, shakes it within his bill, and at the same time utters a series of sounds not unlike those produced by the rattling of grain within his bill. I think the idea is supported by the squealing of the young of the predacious butcher-bird, by various sharp tones of berry- or seed-eating finches, and by some of the sounds uttered by other kinds of birds. There is evidence that some of the song-notes of birds have been developed in the way just mentioned to a semblance of sounds with which the birds are familiar. And in approaching this subject we must remember it is in song that birds first betray most clearly and often their inclination towards mimicry. Some observations subsequently mentioned in detail will prove that several species repeat not only their parents' notes, but also those of other birds around them. There is no ground for supposing that this is any new feature in the economy of birds ; on the contrary, it has probably existed for a vast period of time. If it has so existed—and many birds in captivity will imitate, as we have seen, any prominent sounds near them—is it not possible that during a long period many wild birds have unconsciously learnt to imitate sounds persistent in their neighbourhood, and that these sounds have been, partly by filial mimicry, partly by involuntary imitation, gradually reproduced in their songs? If naturalists in different parts of the world were to make observations in relation to this matter, I feel sure that many facts confirmatory of my views would be established.

Resemblances to Sounds Produced by the Elements

In describing the notes of the American marsh wren (*Certhia palustris*), Wilson stated that, "Standing on the reedy borders of the Schuylkill or Delaware in the month of June, you hear a low, crackling sound something similar to that produced by air-bubbles forcing their way through mud or boggy ground when trod upon ; this is the song of the marsh wren." The belted kingfisher (*Ceryle alcyon*), which lives by running rivulets, waterfalls, and milldams, when startled from its perch, utters a loud, harsh, and grating cry, similar to the interrupted creakings of a watchman's rattle, and almost, as it were, the vocal counterpart to the watery tumult amidst which it usually resides (Nuttall's *Manual*, ed. Chamberlain, vol. i. p. 462, per Mr. A. H. Macpherson).

Several birds of the duck family have whistling cries. The shelduck whistles, as I have observed. So does the wigeon. The summer duck cries *peet peet* (Wilson, *op. cit.* p. 322). The velvet scoter has short squeaking notes (*ibid.* p. 218). The whooper has a whistling cry (Yarrell), and, like the mute swan and the mallard, the sound of its wings in flight can be heard at some distance. Birds accustomed to long flights, and to a life in lonely places, would observe sounds caused by flight, especially at the times of migration.

The voices of mallards, pelicans, flamingoes, and herons resemble the croaking of frogs and toads.

The cry of our common swift has been well suggested in the words *swee ree* (Harting, *Birds of Middlesex*, p. 128); and this, although used by the young in the nest and when chasing their parents for food, is most noticeably employed by the adults on a particular occasion—when they are pursuing each other, or are flying at a great pace together, and when the swish of their wings can be clearly heard many yards below them. It is to be noted that the swift of South Africa "does not pursue and scream like those of Europe" (Layard, *op. cit.* p. 50), and neither swallow nor martin have either habit or cry. In Canada neither the common swift nor the chimney swift, which closely resembles the British species, cries *swee ree*, nor any sound approaching that tone ; nor do they pursue each other in the manner of our swift.

The voices of owls simulate the moaning of the wind in hollow trees, such as these birds frequent.

Sometimes, when in a wood during a heavy storm, I have noticed that the patter of drops of water on fallen leaves was, in rhythm, not unlike the *lit it it* cries of the robin. When walking in the forest-park at Vancouver, B.C., in July, I seemed suddenly to be close to a noisy trickling stream, the sound of which was clearly audible through the dense growth of trees. In a few seconds the sounds ceased, but were soon repeated in exactly the same way. I was astonished to find that the sounds were the song of a little bird, apparently a wren. In that mountainous district rushing streams of all sizes are prevalent. It seems to me that some song-birds, such as the robin, wren, hedge-sparrow, blackbird, and blackcap, which sing mellow tones and intervals of pitch rather than imitations of other sounds, may have acquired this music partly through the influence of the murmurs and gurgles of rippling streams. Most of us have observed the musical sounds caused by small jets of water falling or dripping into water-butts, and some may have noticed the similarly pleasing sounds occasionally repeated almost incessantly and without variation by tiny waterfalls in a streamlet. Having often observed this "music of streams," as it may be called, I one day set down in musical notation as much as possible of the music of a babbling stream. Of course the intervals written could only be considered as representing approximately the sounds produced. Much of this music was far too intricate and vague to follow, and it was only by close attention that I could work out the following examples, which labour was greatly simplified by the continual repetition of the same strains, as commonly occurs. The first was very distinct, and ran thus:

There was a great accent on the note which I have placed first in the bar. Some twenty yards off there was another very musical waterfall, which rippled this:

and so on. The lower notes were the fullest toned. A third had the following:

There are vast numbers of rivulets always uttering sweeter and simpler sounds than these; and throughout the long year, as through the geological epochs that have been, this varying but ceaseless music is continued. Musical ripples are most frequent where a stream runs under bushes, because in such a spot twigs and leaves fall into the water, causing little barriers over which it must mount and tumble on its way to the sea. Many of the warbling birds, such as the blackcap, wren, robin, blackbird, thrush, and willow-warbler, like to build their nests not far from water, possibly because damp spots generally furnish the most perpetual supply of insect food; and in these situations they are often within hearing of some boisterous, or musical little cataract, whose persistent notes may fairly be expected to influence the future songs of nestlings reared in the neighbouring shade. For, in view of the influences which may affect their song in a cage, can we come to any conclusion other than that they would be affected by, and would tend to reproduce distinct musical murmurs of a brooklet close at hand? The song of the British Columbian wren, before mentioned, seems to prove, almost incontestably, that such an original has been imitated by that bird.

RESEMBLANCES TO SOUNDS PRODUCED BY INSECTS

The sounds made by insects are infinitely less persistent than the sounds made by streams, and hence they would be less likely to influence the notes of imitative birds. Wilson stated that the American field-sparrow (*F. pusilla*) has "no song, but a kind of chirruping not much different from the chirruping of a cricket." The bird seems to be partial to thickets (*op. cit.* vol. i. p. 266). Gilbert White observed that the *jar* of the nightjar resembled the *jar* of the mole cricket (Miscell. Observ. in Jesse's *Gleanings*, p. 286). The song of the grasshopper-warbler is exactly like the persistent song of the green field cricket (*Acrida viridissima*), and although the insect is not heard much before August, the young birds must have plenty of opportunities of noticing this strain. Let any one listen to the noise made by our common grasshoppers, and compare the sounds with those uttered as a song by the yellow and cirl buntings, and he must admit that in relation to "time" there is much resemblance between these sounds, and that the tones are not wholly dissimilar. In summer the young buntings must often be in a very din of grasshoppers, for they are generally to be seen where these insects abound, and this is especially the case along sunny hedgerows. From August until the middle of October our hedges are populated with crickets, whose monotonous chirping may be heard both by day and by night. The common call-note of the brown wren is uttered with especial frequency at this season, and it is very closely like this sound, though much louder. Wrens must then hear this chirp throughout the day.

Resemblances to Sounds made by Quadrupeds

The note of the American red-headed woodpecker (*Picus erythrocephalus*) is "shrill and lively, and so much resembles that of a species of tree-frog which frequents the same tree, that it is sometimes difficult to distinguish the one from the other" (Wilson, *op. cit.* vol. i. p. 148). It has been stated that the voice of the ostrich is a roar so like that of the lion that even Hottentots have been sometimes unable to discriminate between them. The sounds are certainly somewhat similar. The cry of the ostrich, like that of the Australian emu, is a deep, hollow, rumbling sound. Wilson said that the note of the American burrowing owl (*Speotyto cunicularia*) "is strikingly similar to the cry of its companion the marmot, and sounds like *cheh cheh*, pronounced several times in rapid succession; and were it not that the burrowing owls of the West Indies, where no marmots exist, utter the same sounds, it might be inferred that the marmot was the unconscious tutor to the young owl; this cry is only uttered as the bird begins its flight" (*op. cit.* vol. iii. p. 225). The South African jackal buzzard (*Buteo jackal*) has a cry "singularly like that of the common jackal, whence its name" (Layard, *op. cit.* p. 27). Verreaux's eagle owl (*Bubo lacteus*) utters a "call-note which is often mistaken for that of the leopard" (*op. cit.* p. 71). Persons who have not been near cattle grazing rank grass at night will think it absurd to compare the sounds thereby produced with the note of the landrail, but the comparison is a very fair one. Cattle, when grazing, protrude the tongue from each side of the mouth alternately, each time round a bunch of grass, which is torn off. A pause then occurs during the act of swallowing, after which the tearing is resumed. The consequence is a sound which may be thus expressed, *rasp rasp—rasp rasp—rasp rasp*. Cattle graze more at night than in the daytime, and the landrail is to a great extent nocturnal; and although the feeding-grounds of cattle are not now the nesting-places of the bird, it is frequently within hearing of grazing, the sound of which is particularly noticeable at night, and has been repeated through probably as long a period as that in which birds of any kind have existed on the earth.

It may here be mentioned that the common squirrel and the snake respectively reproduce in their alarm-cries the sounds made by these animals during rapid retreat. The former utters a sound something like the word *whonk*, but more like the noise produced by swishing with a long twig. The latter produces a hissing sound; and wherever it goes it must always hear the rustling sounds produced by its transit. When it darts through dry grass a regular *hiss* is thus caused. It appears that serpents which live among rocks do not hiss so volubly as those of grassy places.

RESEMBLANCES TO SOUNDS UTTERED BY BIRDS

There are more resemblances between the cries of birds than between those of birds and other animals; and this seems only natural. But here also we find abundant traces of voluntary mimicry, so that we can hardly say certainly whether an individual bird is intentionally reproducing the notes of another, or is repeating the mimicry of some ancestor, descended to him by the agency of filial imitation. It is very possible that birds of different genera, living together, may learn cries from each other. The golden-crested wren, for example, has a call-note which resembles that of the tree-creeper (as Yarrell observed, *op. cit.* vol. iv. p. 470); and these two species are often together. The nuthatch often repeats as a call-note a slight squeak like that of the creeper and goldcrest, and also closely like that common to the coal, blue, and great titmice. In fact, all of these birds, so frequently associated, utter this call. The nuthatch evidently possesses some power of mimicry, for I have heard it repeat exactly the *pim im imimim* of the blue tit. The hedge-accentor is said to sing somewhat like the wren, as it does, but it is as possible that this similarity may have been derived from some persistent source, say, the murmuring of a stream, as that the one bird copied the other. The adherence of this bird to local variations in song can only be accounted for as due to the power of filial mimicry. The Cape broad - billed fly - catcher (*Platyrhynchus Capensis*) " has a curious harsh, loud, and monotonous note which almost exactly resembles that of the owl *Scops Capensis*, and is uttered in about the same intervals, four or five times in a minute" (Layard, *op. cit.* p. 344). The influence of surroundings, in presenting to the notice of birds a restricted range of subjects for imitation, is indicated in their songs. Bechstein wrote of the (wild) red-backed shrike, or butcher-bird (*Collurio*), that " it almost exclusively imitates the birds of its immediate neighbourhood," and that in the house its song is composed of the warbling of the birds hung near (*op. cit.* p. 137). The resemblance of the song of the scarlet bullfinch (*Pyrrhula erythrina*) to " some of the notes of the reed-bunting is a remarkable fact. Both these birds live in its immediate neighbourhood" (Naumann, quoted in Bree's *Birds of*

Europe, vol. iii. p. 72). Bechstein wrote (*op. cit.* p. 254) that the redstart "can improve its song . . . by adding to it parts of the songs of birds that are found near it. For instance, those that build under my roof imitated tolerably the chaffinch that hangs in a cage at my window; and a neighbour of mine has in his garden one that repeats some strains of a blackcap that has its nest near." He adds: "This facility in appropriating the song of other birds is heard in few birds that have their liberty, and seems peculiar to this species."

This last remark is most inapt. The observers here mentioned noticed the mimicry of redstarts in their gardens, presumably because they were well acquainted with the notes reproduced; but Bechstein's remark that few other birds have this power, indicates that he was better acquainted with the notes of caged birds than with those of wild ones. He accurately records that the redstart *adds* the notes of other birds, for the imitations of this species are uttered at the end of the ordinary strain, to which they form a kind of *sotto voce* suffix.

Before going further, it will be as well to make some observation on the difficulty of perceiving imitations sung by wild birds. These imitations can only be detected by a listener who has an accurate memory for the sounds mimicked, and a possesser of this faculty will the more easily perceive imperfect imitations in proportion to its power.

But it is impossible for the listener educated in these matters to teach his knowledge to other people: it can only be gained by close personal attention to the subject, by identifying birds when they exclaim or sing, and by a sufficient power of memory. If, therefore, a person who has studied the voices of birds should claim to have heard certain imitations, the reader must form his own estimate of the statements, which cannot, except in very rare instances, be subjected to proof; but when several well-known authorities have stated (as already quoted) that they had observed similar incidents, the statements must be considered as so much the more acceptable.

It should be observed in this place that some seeming imitations are difficult to relate to one particular bird, as they are like a cry common to two or more species. Thus, as has been noticed, the little upwardly-slurred whistle which is the alarm-cry of both chiffchaff and willow-warbler, and is uttered more loudly by the nightingale, cannot, when imitated, be related to either of these species; for this reason I have adapted a name for the cry, and have called it "warblers' *tewy*." An imitative bird will often mimic two or more cries of others in one phrase: I have taken no regular account of this, nor of the number of phrases betraying no mimicry. The latter are heard in about the ratio of one to three in phrases of the thrush and robin, and nearly one in two of those of the redstart, while there are practically none in those of the starling and sedge-warbler. The degree of practice performed—and this is greatly dependent upon the state of the weather and the abundance of food—has much influence on the fluency of song and mimicry: after prolonged and severe frosts I have heard most feeble and limited songs, as well from starling as from robin and thrush.

THE THRUSH

During fourteen months I listened to an aggregate number, approximately, of 50 thrushes, and I heard 1120 phrases [1] which seemingly contained some recognisable imitation, and 450 in which I detected none. The six subjects most frequently

---

[1] As already stated, the term "phrase" does not refer to any particular strain or order of cries. A bird might repeat the same strain many times, with alternate pauses, and each utterance would be considered a phrase, and recorded as such.

mentioned in the record of these imitations are as follows :—

Bird imitated.	Total instances of this imitation.	Thrushes which imitated it.
Nuthatch [1]	136	37
Wood-warbler	133	37
House-sparrow	103	36
Blackbird's alarm	78	31
Blue Tit	57	24
Great Tit	47	24

THE ROBIN

Pursuing the same method of investigation with the robin, I recorded in the same period 1316 phrases, sung by a total number of 65 birds, and 905 of the phrases contained one or more imitations. The following table indicates particulars of these songs, corresponding to those above stated in relation to the thrush :—

Bird imitated.	Total instances.	Robins which imitated it.
Blackbird's song, or its alarm	132	47
Coal Titmouse	89	44
Hedge-accentor	75	42
Greenfinches' cries	79	39
Lark (calls or song)	99	36
Blackcap	57	23

[1] The tables do not record a sufficient number of imitations of notes of the nuthatch. In early spring I often passed, as not worth notice, thrushes whose notes then consisted almost entirely of modifications of the commoner cries of this bird; and similarly, I have sometimes at first mistaken a loquacious nuthatch for a thrush.

The Skylark

During the same period the lark sang some kind of an imitation 345 times; the most frequent subjects mimicked being as follows:—

Bird imitated.	Total instances.	Birds which imitated it.
Cirl-bunting, or Yellow Bunting	53	31
Swallow	30	20
Tree-pipit	29	19
Blackbird's alarm . . .	21	15
House-sparrow . . .	23	14
Peewit	21	13

The Starling

In the same period I recorded 275 instances of some imitation being sung by a starling, of which species only 18 birds were observed. Their six favourite subjects appeared to be these:—

Bird imitated.	Total instances.	Starlings which repeated the cry.
House-sparrow . . .	36	14
Greenfinch	22	10
Yellow or Cirl Bunting's song .	15	10
Partridge	29	9
Green Woodpecker . . .	16	9
Jackdaw	22	8

The Sedge-Warbler

A total number of 281 instances of imitation given by nine sedge-warblers heard during the same period revealed their preferences as follows:—

Bird imitated.	Total instances.	Sedge-warblers imitating the cry.
House-sparrow	42	9
Chaffinch	24	8
Starling	20	8
Blackbirds' alarm	19	7
Wagtails' cry	18	6
Swallow	21	5

The tables above set out appeared in *The Zoologist* for July 1890. In them the "tree-pipit" appears as the "meadow-pipit," an error not due to inaccuracy in identifying the note, but in identifying the bird. I credited the statements of others in regard to this matter. The meadow-pipit is not very frequent near Stroud, but every year several pairs nest in the vicinity.

All the thrushes, robins, starlings, larks, and sedge-warblers, whose imitations are mentioned in the above tables, were heard within ten miles of Stroud, in which neighbourhood all the birds imitated abound. The nuthatch frequents the meadow elms, and is vocal throughout late autumn, winter, and spring; and during these seasons the voices of the great and blue titmice are very noticeable. In gloomy dells about the wooded hills the coal-tit is very common; and here young robins are to be found during summer and early autumn—at which period, as we know, they are receiving strong impressions of sounds. The hedge-accentor and blackcap are very common, especially in the haunts of the robin. In the open fields we find buntings numerous, and during summer the swallow's voice is heard almost as often as theirs; while in every sheltered valley, and up the hillsides, the tree-pipit pours out his music. The peewit frequents the open fields throughout the year, but in small numbers. Greenfinches are abundant. Every evening the partridge's call may be heard in the fields. The jackdaw is plentiful, and is often associated with rooks and starlings. The chaffinch is abundant, as are also the commoner wagtails in their seasons.

Thus we find that each of the five imitative species above mentioned has preferred to reproduce the cries of birds which are abundant, or whose voices are especially noticeable, in its neighbourhood.

Song-birds which are imitative are not unnaturally influenced by changes in the more persistent sounds that are heard around them. The following table, for which no selection of particular singers was made, indicates changes induced in the song of the robin by the songs of birds which are summer visitors to Great Britain. In this table the word "chaffinch" denotes all the cries of the chaffinch, some of which are often uttered in winter. It may be as well to mention that the wood-warbler, blackcap, and willow-warbler are all common summer visitors to the district (Gloucestershire) where the robins in question were heard.

From Sept. 1887 to 1st April 1888—		From 1st April 1888 to August 1888—	
Number of robins	. . 50	Number of robins .	. . 16
Total of phrases recorded in this period	. . . 890	Total of phrases recorded in this period	. . . 426

Bird imitated.	Percentage of imitations in total of phrases.	Percentage of imitations in total of phrases.	Extent of change, per cent.
Chaffinch	4	1.1	2.9 decrease
Lark	7.9	5.3	2.6 ,,
Accentor	6	4.9	1.1 ,,
Wood-warbler	2.9	4.9	2 increase
Blackcap	3.3	6.3	3 ,,
Willow-warbler	.6	5.8	5.2 ,,

Here we have records of imitations of two winter and early spring songsters (lark and accentor) and the common chaffinch, being to some extent supplanted by those of newly-arrived warblers. These examples are the most pronounced: many of the cries of common birds are imitated almost equally throughout the year.

Mimicry of the Thrush

In comparing my records of the songs of the thrush heard in the first months of the year with those heard in May and June, a change of song, similar to that shown in this table, is evident. I find that in the latter period the blackbird's alarm is, proportionately with the other imitations, much less often uttered; while the reproduced notes of the cuckoo, wood-warbler, and butcher-bird (or the note of the wryneck) then become much more frequent.

I have made a list of the subjects which I have recorded as sung by all the thrushes (approximately about 70 in number) heard by me in Gloucestershire. It is possible that some of these birds may have been counted twice over, but I exercised the greatest care to avoid such an accident. The table shows how many of each kind of imitation were exact reproductions; but the number of these must have been greater than appears, because I was not so careful as I might have been in regard to this point, and sometimes forgot to note whether the reproductions were exact. In fact, throughout the ensuing tables the imitations numbered as exact must be read as being extremely well performed by the respective singers; for some birds, such as the starling and sedge-warbler, are habitually very accurate in their mimicry.

Subject imitated.	Thrushes.	Times imitated.	Exact reproductions.
Nuthatch	50	214	5
Blackbird's alarm	44	116 ⎫ 125	1
Blackbird's song	9	9 ⎭	...
Crow	34	107	2
Cuckoo	21	78	4
Coal Tit	31	64	1
Great Tit	30	62	...
House-sparrow (undescribed)	22	51 ⎫	...
(autumn call) *whceo*	5	7 ⎬ 98	...
tell tell	14	17 ⎮	...
chissick	12	23 ⎭	1
Greenfinch *yell yell*	16	19 ⎫	...
titititit	11	22 ⎮	...
(undescribed)	8	15 ⎬ 68	...
upwardly-slurred alarm [1]	4	9 ⎮	...
final song-note	3	3 ⎭	...
Green Woodpecker	24	51	1
Cry of chick	22	53	3
Partridge's call	17	41	3
Wood-warbler	17	46	6
Butcher-bird	20	33	2

[1] This note is an upwardly-slurred whistle, similar to, but more prolonged than the alarm called warblers' *tewy*. I have heard it commonly uttered by hen greenfinches disturbed by some rapacious bird or beast near the nest. It is also uttered by the redpoll.

Subject imitated.	Thrushes.	Times imitated.	Exact reproductions.
Mistle-thrush	17	28	2
Tree-pipit	16	28	7
Corn-crake	16	28	1
Nightingale	10	28	...
Starling (whistle)	18	22	...
Sedge-warbler	9	22	1
Wren's call-note	19	21	...
,, song	...	1	...
Lark	11	21	...
Warblers' *tewy*	9	21	...
Chaffinch, song	4	4 ⎫	1
fink	9	11 ⎬ 18	...
loud *twit*	1	3 ⎭	...
Yellow or Cirl Bunting's song	8	8 ⎫ 13	...
Yellow Bunting's single cry	5	5 ⎭	...
Whitethroat	11	15	1
Swallow	10	12	1
Moorhen	6	13	4
Wagtail	5	8	...
Blue Tit	9	10	...
Goldfinch	8	9	2

There were many other apparent imitations less frequently heard. The cry called *be quick*, often heard in the song of the thrush, and which blends into imitations of the common call of the nuthatch and the cry of the guinea-fowl, are not mentioned in this table. This was sung much more than the 115 times I have recorded as noted in an aggregate of 29 thrushes. The reader must not suppose that birds can imitate all sounds with equal ease; on the contrary, their voices are often quite unsuited for the mimicry attempted. The cry of the crow (a common bird near Stroud) affords an example. I have a note that on 21st March 1890 I heard many thrushes imitate the crow, at Sans' Wood, where it abounds. On the 14th March I heard a thrush in my garden exactly reproduce a cry which I had heard on the previous night repeatedly and loudly uttered by some strange bird flying overhead. I never again heard this note, which was a remarkable one. In the same month I heard a thrush imitate the crow of a bantam exactly. Another exactly reproduced the clucking of a hen for her chickens. Both birds were close to farm-buildings where scores of fowls are kept. The clucking of a hen I have also heard exactly reproduced by a thrush at Brimscombe; this was in June. The following is an example of my records. A phrase without recognisable mimicry is indicated by an "O."

"Frocester, Glos., near the church, 17th May 1892.—Thrush singing:—golden plover—golden plover, O—crow—corn-crake—be quick, O, O—wood-warbler's sibilous notes—cuckoo (in rough tones), O—young starling's cry after leaving nest, O—

butcher-bird—be quick, O, O—whitethroats' alarm—
great tit (cry), O, O—end."[1] The cry named after
the "golden plover" occurs elsewhere: I so name
it on the authority of a friend. I have never seen
the bird which utters it, having heard it at night;
but I know it as well as I know the bark of my
brother's dog—a constant companion. On 8th
July 1891 I heard a thrush at Stratford Park,
Stroud, exactly reproduce the song of a neighbouring
tree-pipit, even to the exact pitch of the notes.

The thrush often reproduces the musical intervals
uttered by one bird, but in the tone of another.
Thus, it will whistle notes in the interval of a
third, like a cuckoo, or it may utter them in a
rough voice like that employed to imitate the
crow. There seems to be no end to the versatility
of the thrush in this particular. Sometimes, also,
it will associate certain imitations and repeat them
several times in the same order. Thus, at Bath,
on 3rd April 1892, a thrush was persistent in
reproducing the hawk-alarm of the house-sparrow
(a sound somewhat like the word *tour*), immediately
followed by the cry *tell* of the same bird; the
result was a prolonged repetition of *tour tell, tour*

---

[1] I found a knowledge of phonography very useful for recording songs, on account of the pace at which a record may by this means be made.

tell, etc., the accent being on the second syllable. I must repeat, that imitation, like variation, is most in evidence towards the end of the season of song. In March 1889 I was staying at Weston-super-Mare, at which place the thrushes abundantly imitated the dunlin, there a common and noisy bird throughout the colder months. At that time the dunlins could be heard all day long uttering their call-notes. I have a note, that on 10th March a thrush in a garden in front of the house (Manilla Crescent) had been imitating the dunlin so much that I thought a flock of the birds must be in the vicinity, until I watched the thrush singing within the distance of a few yards. It rarely uttered the cry of the dunlin without repeating it many times, just as it is heard when a flock of dunlins is near. Here is an extract from my notes, recording a song by this thrush:—
" Dunlin—dunlin—wood-warbler's plaintive note (alarm)—dunlin—house-sparrow's *chissick*—crow—crow—dunlin—crow—crow—tree-pipit—dunlin—dunlin—dunlin—dunlin—crow, O—crow—bequick."
Each of the above subjects represents a phrase. Out of 37 phrases sung on the same day by a thrush near Worle, three miles inland, eight contained imitations of the dunlin. In *The Zoologist*

for 1867, p. 599, Mr. W. Jeffery, jun., statest hat near the sea the song-thrush and skylark vary their songs occasionally with the ringed plover's whistle (*vide* Mr. Harting's note, p. 209 *post*).

MIMICRY OF THE ROBIN

I have made an epitome of the principal imitations sung by the robin, similar to that above given in regard to the thrush. This relates to nearly seventy robins heard during the last four or five years in Gloucestershire; and in this the exact imitations are set out, in consequence of their having been so much more noticeable in this species, which cannot be termed generally a good mimic.

Subject imitated.	Robins.	Times reproduced.	Exact reproductions.
Blackbird's song or Mistle-thrush's song	26	43	7
Blackbird's alarm	45	105	15
Coal Titmouse	55	120	14
Lark	43	113	13
Greenfinch	48	103	6
Hedge-accentor	44	76	14
Blackcap	39	72	16
Chaffinch	35	53	8
Great Tit	33	45	4
Thrush	28	40	4
Wood-warbler	33	40	2

Subject imitated.	Robins.	Times reproduced.	Exact reproductions.
House-sparrow	25	34	3
Blue Tit	19	29	5
Swallow	11	28	3
Yellow Bunting	19	22	5
Willow-warbler	8	10	1
Common Bunting	7	9	...
Starling's whistle	8	8	4
Chiffchaff	6	9	2
Nightingale	7	8	...
Green Woodpecker	7	7	1
Whitethroat	7	7	1
Goldfinch	6	6	1

The robin often associates a number of imitations in one phrase. I once heard six different (and well-imitated) cries of other species uttered in unbroken succession by a robin; but as a rule mimicry is replaced by the peculiar trickling music characteristic of this bird. As in other species, there is considerable variation in the duration of the pauses occurring between the phrases, the latter being longest when the best singing is attempted. The memory of the robin is quite equal to the retention of an accurate idea of a song six months after it has been heard; for, in January, one may often hear a robin reproduce exactly the song of a black-cap, that of the willow-warbler, or the coarse final note of the greenfinch. Yet, even in regard to this faculty, individuals seem to differ in ability, and I have a note that, on 21st March 1890, I listened during seven minutes to two robins singing in a thicket where willow-warblers are most abundant, and I heard only one imitation of this bird's song; yet in summer I have heard it frequently mimicked by robins at the same spot. On 15th February 1891, I heard at Grove House, Painswick, a robin reproducing exactly the double alarm-cry, *tewy chick*, of the redstart; and I should have been deceived had I not known that the redstart rarely utters this cry except when the young are about, and so been induced to look for the singer. I saw the robin singing, but no redstart, nor have I ever seen one of the latter in winter or before the end of March.

MIMICRY OF THE SKYLARK

The song of the skylark is equally interesting as revealing a reproduction of cries of neighbouring birds. Here is a summary of the principal imitations sung by some thirty larks :—

Subject imitated.	Skylarks.	Times reproduced.	Exact reproductions.
Yellow Bunting's song	26	66	5
Tree-pipit	17	28	...
Swallow	18	26	1

Subject imitated.	Skylarks.	Times reproduced.	Exact reproductions.
Blackbird's alarm	14	22 } 24	1
,, song	2	2	...
Martin	15	21	1
House-sparrow	14	21	...
Chaffinch	8	21	2
Peewit	11	19	3
Green Woodpecker	11	12	...
Wagtails (inclusive)	10	17	...
Common Bunting (song)	8	14	...
Hedge-accentor	8	11	...
Greenfinch	8	8	...
Sheep's bleat	7	9	...
Magpie	6	7	...
Whitethroat's alarm	5	8	...

It should be observed that on the hills near Stroud, where larks abound, the magpie is a common bird. The "sheep's bleat" above mentioned is difficult to identify in so musical a voice as that of the skylark; it is a short but distinct sound, only to be described in the word *baa*.

Mr. Harting writes: "I have frequently noted in Sussex, Essex, and Norfolk, that in marshes near the coast, where the skylarks have constant opportunity of hearing the notes of the ringed plover, they, either consciously or unconsciously, imitate them so perfectly that I have been momentarily puzzled to know which species was calling" (*in litt.*).

(See also *The Zoologist*, 1867, p. 599.) Near Weston-super-Mare I heard the larks occasionally imitating the cry of the dunlin, which bird abounds there in winter.

While crossing Salisbury Plain, from Salisbury to Devizes, on 3rd June 1892, I heard scores of larks in full song. I was on the plain for four hours, and the weather being bright, I had abundant opportunities for making observations. Nearly all the songs of larks heard there by me were composed of repetitions of the common call-notes of the species (which may be suggested by *worryou* and *teeuu*), which were uttered in set orders of repetition by different individuals. One bird imitated the house-sparrow's *tell tell* many times, also the cry of the wagtail a few times. Another lark imitated the yellow bunting's song, and then uttered a low rattle which I could not refer to any bird.

Many of the larks imitated the cry of the peewit; in fact, their songs consisted almost exclusively of their own call-notes and cries of the common and yellow buntings and the peewit. I never heard one imitation of the robin or blackbird, or of a titmouse. But as I approached Chippenham, on the same day, I again heard these familiar imitations; and one song, heard six miles from Devizes, is here transcribed:—Blackbird's alarm—martin's cry, exactly—yellow bunting's song—chaffinch's *fink fink*, exactly—swallow's song. Every imitation was a very good one. Mr. Sutton A. Davies, of Winchester College, informs me (*in litt.*) that the larks in his neighbourhood imitate the common bunting.

MIMICRY OF THE STARLING

The following is a similar summary of the principal songs of starlings heard near Stroud:—

Subject imitated.	Starlings.	Times reproduced.	Exact reproductions.
House-sparrow	21	67	5
Jackdaw	14	37	...
Greenfinch	14	28	...
Green Woodpecker	14	23	1
Blackbird's alarm	13	20	1
Partridge	11	20	...
Chaffinch	10	20	...
Nightjar	10	19	...
Swallow	10	16	1
Yellow Bunting's song	8	13	...
,, ,, call	8	11	...
Blackbird or Mistle-thrush's song	8	13	1
Thrush (alarm)	7	11	...
Peewit	5	8	...
Dog's bark	5	7	...
Nuthatch	4	7	...
Migrating Redwing's cry	5	7	...
Robin (alarm)	5	8	...
Blue Tit	5	6	...

It is practically unnecessary to distinguish the varying degrees of accuracy in the imitations of the starling, for this bird is one of the best mimics, and its reproductions of the notes of other birds, and even of animals, are as exact as they are various. I often hear a starling imitate the call-note of the tawny owl, which bird is common in this neighbourhood. The starling can imitate another bird and at the same time utter quite different sounds: it is difficult to imagine how he accomplishes this feat, but the fact remains patent to close observation. The sounds which I have named as imitations of the nightjar are made with the mandibles, which apparently are rattled for this purpose. A starling which imitated the cries of a hen is recorded by Mr. L. Buttress, of Grove Rectory, Retford (*The Field*, No. 2054, p. 666). On 2nd March 1892, at Stratford Park, Stroud, three starlings imitated exactly a peculiar *swurrr*—often uttered by some water-fowl about the pond in the park.

Mr. A. H. Macpherson has kindly sent me the following note:—" In 1887, at Trinity College, Oxford, I heard a starling on the roof at the opposite side of the Quadrangle attempting to imitate the chapel bell, which was then ringing. To my surprise I noticed that, in addition to imitating the sound, it was swaying its whole body from side to side in imitation of the movement of the bell."

The late Mr. J. E. Anderson of Lilswood, Hexham, wrote to me as follows: "I have several times gone to the door to let in a cat—it was only a starling practising mewing! They are very fond of imitating the curlews, and do it well as far as they go.[1] One of them was bothered the other night in attempting to imitate the call of the partridge: he failed miserably. There are two colonies of starlings here. One bird claims the chimney—that is the one that imitates the curlews, cats, and birds of notes dissimilar to his own. Another claims some trees close to where he builds: his imitations are confined to blackbirds, and a few notes from the lark, mixed up in a very hearty manner with his own song. Sometimes he is quite jubilant; and, did he not change from 'blackbird' to 'starling,' I could not distinguish him from a blackbird. He hears the blackbirds in the adjoining plantation."

The starlings near Stroud can generally imitate

---

[1] Mr. Harting informs me that when staying on a visit at Vagnol Park, Bangor, close to the Menai Straits, where curlews are often seen, and come into the park to rest at high tide, he was much struck with a starling which used to sing every morning near the house, and give an admirable imitation of the curlew's note.

the partridge with facility; I have, however, often seen them evidently exerting themselves in producing the cry of another bird; one in particular seemed compelled to completely extend his wings when imitating the cry of the peewit.

MIMICRY OF THE SEDGE-WARBLER

I have made records of the songs of about fourteen sedge-warblers heard near Stroud. The following were the principal imitations:—

Subject imitated.	Sedge-warblers.	Times reproduced.	Exact reproductions.
House-sparrow's *chissick*	11	31 ⎫ 65	6
tell tell	9	34 ⎭	...
Brown Wren's call	8	36 ⎫ 43	...
,, ,, song	2	7 ⎭	1
Starling	11	23	...
Warblers' *tewy*	9	19	1
Swallow	8	39	1
Wagtails	8	21	2
Chaffinch *fink*	6	20 ⎫ 29	...
,, *twit*	6	9 ⎭	...
Blue Tit	7	15	2
Butcher-bird	7	12	1
Martin	6	15	...
Corn-crake	5	15	1

The imitations sung by the sedge-warbler are generally remarkable for correctness; but some of them are yet superior to the others, and of such I have noticed not a few. This bird generally strings together many imitations in one phrase, the song being usually commenced with the ordinary notes, *jig jig*, and afterwards varied in mimicry. The sedge-warbler (like the lark, robin, and thrush) repeats these imitations with varying degrees of rapidity, and in a repeated order of succession, so that it may be said to construct its own strains out of the songs of other birds. On 4th June 1892, near Crudwell, Glos., I heard a sedge-warbler uttering one of these original phrases again and again. It began slowly, *jig jig*, followed by an exact reproduction of the *chissick* of the house-sparrow. Thus the strain ran, *jig jig chissick*, and so on; only the pace was greatly increased as the phrase proceeded; and this was concluded with ordinary imitations. It also sang *jig fink*, the second note being exactly that of the chaffinch; and in this case also the repetition was accelerated as it progressed. In precisely the same manner, the bird repeated the *tell tell* of the house-sparrow with the *fink fink* of the chaffinch, and the common alarm-cry (*clittit*) of the swallow with the *jig jig* of the sedge-warbler. This individual is mentioned as especially exemplifying a habit general in the species. At Lower Dorington Terrace, Stroud, a caged lark has been kept for many years, and the bird has learned the calls of canaries which are kept by a neighbour, also the call of the brown wren, some notes of the great tit, tree-pipit, and yellow bunting, which notes it repeats in an almost incessant song. I have heard a sedge-warbler in a marsh, about 120 yards distant from, but within easy hearing of the lark, singing phrases closely like those of that bird.

The most remarkable song of the sedge-warbler noted by me was uttered at ten o'clock at night, near Chalford. A bird of this species was singing on the other side of a roadside wall, and some few feet down a bank. As I looked over the wall the song ceased; and I waited, watching the bush whence the sounds had proceeded. Suddenly I heard the cry of a chaffinch—the call employed in flight—so loud, and yet so rapidly diminishing in force as the cry was repeated, that I felt certain a chaffinch had been startled from the bush—an unusual occurrence under the circumstances—and, more strange still, had uttered its call-note as it flew far out over the valley. I prepared to make a note of this, and actually took out my note-book for the purpose. Suddenly I heard precisely the same cries—the loud call fading away—as though a second finch were still calling, though retreating far across the valley. Then I knew that the sedge-bird had deceived me. I then heard it exactly reproduce the twitter of a canary, which, at that time, was hung on a house about fifty yards away; and then it uttered the most wonderful song I ever heard. It sang a long phrase consisting wholly of alarm-cries, such as one hears when a hawk comes over a numerous population of small birds. The vehement double cries of swallows, the loud *cah cah cah* of the starling, the cry *tourr* of the house-sparrow, and its *tell tell*, with the alarm of the blue tit, were jumbled together as they sometimes are in nature; and then the sedge-warbler continued his phrase in the single cries of *tell* in which the old male house-sparrow, watching as a sentinel, warns his neighbourhood that a hawk is very near. All other cries were abandoned when this signal was reproduced; and, with lengthening pauses between these notes, the phrase gradually came to a conclusion. The succeeding phrase was of the ordinary kind. On the following day I went to the spot, and saw the singer, a sedge-warbler, in the same bush.

Mimicry of the Redstart

The following is a summary of the principal songs of five redstarts heard near Stroud, in April and May 1888:—

Subject imitated.	Redstarts.	Total instances.
Brown Wren's song	2	10
,, ,, call	2	2
Willow-warbler	3	5
Whitethroat	3	4
Tree-pipit	2	7
Blackbird's alarm	2	5
Chiffchaff	2	5
Thrush	2	5
Nuthatch	2	4
Coal Titmouse	2	4
Chaffinch	2	4
Nightingale	2	4
House-sparrow	2	3
Greenfinch	3	4
Sedge-warbler	2	3
Swallow	2	2

Mimicry of the Nightingale

In stating that the nightingale had only twenty-four strains, Bechstein recorded his own lack of observation. We can no more define the limits of its song than those of the voice of a thrush or redstart. I remember one which sang so like a whitethroat that a clerk in my employ, who knew but little about birds, noticed the resemblance and mentioned it to me. One I heard exactly repeat the cricket-like chirrups which the chiffchaff so often utters immediately after its own typical cry. One repeated the *chissick* of the house-sparrow fourteen times without a break. As I have before remarked, there is individually much variation in the frequency with which one of the call-notes (an upwardly-slurred whistle) is repeated at the commencement of a phrase. Several reproduced the "water-bubble" song of the nuthatch exactly; and another, the green woodpecker's cry, three times in twenty-eight phrases. I tried to take down the song of one of them, in the method adopted by Bechstein. Here is a part of the record: —*Quee, quee, quee* . . . *Tsorr tsorr* . . . *Peeuu peeuu* . . . *Tso, rrrrrr he. Rrrrrrr se. Whit rrrrrr. Tsu tsu tsu.* . . . Woodwarbler, exactly. House-sparrow's *chissick. Tewy. Pee pee* . . . *ke. Tewy*, and blackbird's alarm. *Highlo highlo* . . . *klo klo klo*, etc. I have several times heard one repeat a phrase which has just been sung by another. The fulness of tone which the nightingale displays interferes with accuracy of imitation in many instances; and indeed, so wonderful is the song, that a listener is apt to forget all else than the supreme impulse and passion of the singer. Perhaps the surroundings of the bird increase this effect. The murmur of a stream; the soft moonlight which sometimes bathes the dewy meadows, and sheds white waves across the road or the woodland track, chequered with shadows of clustering fresh May leaves—these are suitable features in the realm of this monarch of song, and increase his effects. Now he prolongs his repetitions till the wood rings. Now his note seems as soft as a kiss; now it is a loud shout, perchance a threat (*rrrrrr*); now a soft *peeuu, peeuu,* swelled in an amazing crescendo. Now he imitates the *sip sip sip sisisisisi* of the wood-warbler, now the bubbling notes of the nuthatch. The scientific investigator is abashed by this tempestuous song, this wild melody, the triumph-song of Nature herself, piercing beyond the ear, right to the heart of the listener. He is pleading now! But no, he is declamatory; now weird, now fierce; triumphant; half-merry: one seems to hear him chuckle, mock, and defy in almost the same breath.

MIMICRY OF THE MARSH-WARBLER

In *The Zoologist* for August 1892 is a graphic account of the mimicry displayed by this bird,

by Mr. W. Warde Fowler. He heard one at
Meiringen, in the Oberland, Switzerland, which,
"besides the notes of birds, distinctly appeared to
me to take pleasure in imitating the sharpening of
a scythe, a sound frequently to be heard from an
adjoining field." Another deluded Mr. Warde
Fowler into the belief that a chaffinch was singing
in the same bush, and "once I fully believed I heard
the nuthatch's clear metallic note. He also mimicked
the skylark, the great tit, the white wagtail, the tree-
pipit, and the call of the redstart." Mr. Herbert C.
Playne (who subsequently watched marsh-warblers
with Mr. Warde Fowler in an osier-bed in Oxford-
shire) described this species to me as greatly excel-
ling the sedge-warbler or the reed-warbler in mimicry.
He states that the bird sings each imitation separately,
as though thinking a moment before every effort, and
that each of its attempts in this particular is wonderfully successful.

Mimicry of the Reed-Warbler

A reed-warbler heard by me at Brimscombe, near
Stroud, imitated many times the cries of the starling,
including the common cry of alarm (the *cah* employed
as an alarm to the young), and the song of the

starling. A pair of the latter species had a nest within ten yards of the singer; hence I was able to compare the imitations (which were excellent) with their originals. The swallow, wagtail, and house-sparrow were also abundantly imitated. The swallow's song was capitally rendered seven times successively. Mr. H. C. Playne informs me that he has heard numbers of these birds near Oxford, and that they are good mimics.

MIMICRY OF THE WHEATEAR

This bird is not generally considered a good mimic, but the late Dr. H. L. Saxby held the opposite opinion. He wrote: "Upon very many occasions I have heard the wheatear successfully imitating the notes of the following birds:—House-sparrow, skylark (part of song), common bunting, mountain linnet, peewit, golden plover, ringed plover, redshank, oyster-catcher, and herring-gull. So complete is the deception, that when the bird has been out of sight I have many times been thoroughly taken in" (*Birds of Shetland*, p. 68).

MIMICRY OF THE GOLDEN-CRESTED WREN

The song of this diminutive bird generally consists of repetitions of a shrill squeak, accelerated towards the close; but sometimes this strain is supplemented with variations, and even with imitations. On 26th February 1890, I approached to within three feet of a goldcrest singing in a closely-cut hedge. Sometimes it uttered the notes of the coal titmouse, in other phrases those of the blue titmouse, and short squeaks like those of these tits, also the *fink* of the chaffinch. On the 23rd of March following I heard a goldcrest reproduce exactly the *pim im imimim* of the blue tit, and it also sang a note of the coal tit.

MIMICRY OF THE WHITETHROATS

Both the whitethroat and lesser whitethroat often mimic a little, but their imitations are not very numerous nor distinct.

MIMICRY OF THE NUTHATCH

As already stated, I have heard the nuthatch repeat exactly a characteristic note of the blue tit.

MIMICRY OF THE REED-BUNTING

I have heard reed-buntings reproduce the *tell* of the house-sparrow or the greenfinch, and the *fink* of the chaffinch, also another cry of the brown linnet and greenfinch.

Mimicry of the Stonechat

This bird also mimics, but not so well as many other birds. Its song is weak, and its phrases are not longer than those of the redstart. I have heard it many times reproduce nearly the whole of the song of the yellow bunting, also the song of the common bunting, and part of that of the chaffinch. I have only heard a very few of these birds sing.

Mimicry of the Blackbird

The blackbird is about as good a mimic as the blackcap. An observer must be fairly close to either of these birds in order to make out its imitations; for these, in either species, are rarely uttered in the full tones of the ordinary songs. The blackbird has been known to crow like a cock, and flap his wings at the same time. I have heard a blackbird at Tortworth, Glos., imitate repeatedly the cry which I am informed is that of the golden plover. I have also heard these birds reproduce cries of the greenfinch, blackcap, wood-warbler, nuthatch, peewit, swallow, great tit, green woodpecker, goldfinch, and magpie. Mr. F. A. Chambers, of The Elms, Stroud, informs me that when staying with friends at a distance, he was informed that a blackbird in the garden had learned the two first bars of the then very popular song "Two lovely black eyes." Mr. Chambers afterwards saw and heard the bird sing these notes so incessantly as to be wearisome.

MIMICRY OF THE CHAFFINCH

I once heard the greenfinch, but never the house-sparrow, imitate another bird; but I have more than once heard the chaffinch imitate the greenfinch. In 1892, a chaffinch which had its nest in the garden at Brookside, Chalford, exactly reproduced the song of a greenfinch which repeatedly flew around the spot, singing its ordinary flight-song (in which the final note *zshweo* is not uttered); and the chaffinch would sing its own ordinary phrase and then that of its neighbour and relation, with equal facility. I have sometimes heard a chaffinch sing only a small part of the song of the greenfinch. In May 1890, at Dursley, I walked close up to a male chaffinch which had quite deceived me with its utterance of the common call-note of the pied wagtail. The bird seemed to employ this cry for somewhat of the purpose of a call-note, uttering it at intervals, and sometimes interrupting the repetition by a few utterances of its ordinary *fink*. Early in the spring of 1892, I heard at Stroud a chaffinch which uttered this cry of the wagtail with the accuracy of the Dursley bird. It is possible that this cry was a modification of the soft *chissick*, which is a love-call of the chaffinch.

MIMICRY IN OTHER BIRDS

I have heard slight indications of mimicry in the linnet. There are, of course, many birds other than those already recorded which imitate when wild, but the above instances may to some extent indicate the influence which mimicry must exercise in the development of birds' songs. It would seem that cries of one species may be gradually adopted into the songs of other species; indeed, the above lists of imitated subjects reveal this as at least a temporary occurrence. To the effects of this agency may be attributed some of the similarities observable in certain cries of diverse species. The long, full, "water-bubble" phrase of the nightingale has always seemed to me to be closely like the "water-bubble" notes which form one of the most noticeable spring-songs of the nuthatch. Many lesser whitethroats (but not by any means all of them) utter a similar long roll of very rapid, full-toned repetitions of a short note; and the redstart also generally utters towards the commencement of its phrase a much shorter but otherwise somewhat similar repetition. I should not consider this full, loud phrase of the nuthatch as of imitative origin, for the old bird utters to its young and also to its nesting-mate a cry of precisely similar construction, but almost toneless; this is as it were a whispered repetition of the full bubbling song, of which it is probably the original. I believe that young nuthatches, when crying for food to a near parent, utter this toneless repetition.

It is probable that many curious notes uttered by our imitative songsters are survivals of imitations of the cries of species no longer common, and that some others have been acquired from foreign birds. The latter feature would especially occur in the songs of the sedge-warbler and other imitative summer visitors to Britain.

The reader will have perceived that my suggestion that birds' notes, whether cries or songs, have been modulated to resemble sounds with which birds are familiar, is a very rational one. Analogous similarities of form and colour are found in many other animals and insects; and the claim that such features are developments from other types is admitted by most naturalists. But no individual bird has ever been known to radically change its colours, even under the influence of the most complete change in its surroundings; yet, as we have seen, many song-birds exhibit such a change in their songs to an extent patent to all who have paid attention to the subject. Is it strange, then, that a woodpecker should have a cry exactly like the note of its neighbour a tree-frog (see p. 188), whose cries may be a survival of the complainings of the permian epoch?[1] Is it wonderful that in autumn the brown wren should particularly affect a little chirp (not the love-call-note) like the chirp of its companion at that season, a cricket, whose note may have first been produced by an orthopterous ancestor in the coal period? Can we wonder that the young of the imitative butcher-bird, when out of the nest, should squeal like a tortured frog or bird, when we know that the parents slay frogs and birds in the vicinity of the young? It is but natural that the blackbird, which in winter so often hears the curious toneless bubbling sounds audible when one holds a cracked snail close to the ear, should terminate its song with

[1] I shall never forget the loud shrilly noise made by some so-called "tree-frogs" at Ceres, Cape Colony. In March 1885 they produced, every night, quite a din, rendering conversation under the trees a matter of some effort.

similar high toneless noises. Similarly, there is nothing extraordinary in a resemblance between the noise one hears when pulling a large earthworm from its hole (the ear being near the ground) and certain low-pitched "whirry" noises often uttered by the starling towards the commencement of a song.

All evident imitations furnish proof of the influence of surroundings in regard to the evolution of bird-voices. When we remember that probably through long ages this principle has been ceaselessly operating, and that its effects have not necessarily been lost in each generation, but probably have been perpetuated through the agency of filial imitation, we need not wonder that the cries of birds (imitative or unimitative) are so often somewhat similar to the sounds which the birds themselves experience daily, either in relation to obtaining food and to feeding or to the other incidents of their lives. On the contrary, it may be justly surmised that nearly the whole range of bird-song may have been affected by the imitative faculty, which we know to have so wide-spread an influence in the animal world; and that the voice of the bird has been thus attuned to harmony with neighbouring sounds, just as its colours so often blend with those of its surroundings.

CHAPTER X

THE MUSIC OF BIRD-SONG

IT is obvious that many birds of limited voice repeat in their cries various intervals of musical pitch. The cuckoo, cock, chiffchaff, and great tit afford familiar examples of this feature, which is traceable in many songs much more extended than those of these birds, namely, in those of the chaffinch, greenfinch, hedge-accentor, willow-warbler, blackbird, blackcap, brown wren, and others. It is easy to record the intervals expressed in limited cries, though, of course, with only approximate accuracy, for the birds have no knowledge of our scale of music. But when birds sing such long phrases as those uttered by the robin, blackbird, and blackcap in June, it is difficult to make one individual record; and impossible—such is the variety of these songs—to make any true general record, as characteristic of the species; though this has often been attempted, especially in popular journals, by contributors who seemed to imagine that when they had taken down one or two phrases of one, or perhaps of two, individuals, they had obtained a full transcript of the song of a species. I trust that the variety of the music which I have written down (though but fragmentary) may suggest the falsity of such limitations.

Although possessed of a musical ear, which from childhood had been exercised in various musical pursuits, I was at first greatly puzzled in attempts to follow the intricacies of bird-music, which are often executed so rapidly as to be not only difficult to follow, but sometimes actually impossible to record in their natural order. My method of noting the music of birds was as follows:—I did not attempt to write all the music of a rapid singer, but listened for some phrase sufficiently simple for my purpose, and then carefully wrote it down. By this method, slow though it was, I was enabled to obtain records which, although not perhaps scientifically accurate, were as true as musical notation would allow. Many of the phrases of both the thrush and blackbird came wonderfully near, indeed, many seemed identical with, the intervals of our scale. Such an incident does not appear surprising when we consider the imitative powers of the best singers, and the frequency of human music in their haunts. The field-labourer whistles; from villages issue louder, though not always sweeter, musical sounds; throughout the year music is heard in country towns. It appears also that our musical scale is of remote origin, and that for thousands of years the intervals which we now employ have been wafted from musical instruments used by men to the ears of listening birds. On 3rd December 1890, according to the *Daily News* of the following day, Mr. T. L. Southgate delivered to the students of the Royal Academy of Music an address on ancient Egyptian musical instruments, and especially in relation to some flutes 3000 years old, then recently taken from a tomb at Kahun. Mr. Southgate, "admittedly one of the greatest of living authorities upon the subject of ancient musical instruments," proved in his address how abundantly music was employed by the ancient Egyptians. Two of the flutes were in a sufficient state of preservation to be played. "Performed upon by Mr. J. Finn, they gave practically the exact notes of our diatonic scale, thus proving— in every sense of the term to actual demonstration —that our scale was known to the Egyptians many centuries before the Greeks, from whom it had erroneously been supposed we borrowed it."

Mr. Hughes informs me that a parrot in his possession learned to whistle the four notes of the common chord :

and that a starling on the house picked up the strain and sang it with exactly the accent of the parrot, which was often placed out of doors. That birds should repeat a particular phrase in the same successive intervals of pitch is no more strange than that people should do the same thing. Hawkers and railway men are particularly liable to exhibit this tendency. Mutual aid, though so often advantageously employed by birds for the detection of danger, is never otherwise exercised, except in domestic incidents; consequently, birds' cries are largely restricted to the purposes of call-notes and danger-cries, and do not require the elaboration necessary in more highly developed social conditions.

In making my records I have paid no attention to actual pitch—I believe that this has no scientific value—but all the purpose of my records is to suggest intervals between notes sung by birds. In every instance the notation is probably from one to two octaves lower than the actual notes it represents.

In September 1890 I heard a cock utter an unusual crow:

In the following September, at a different place, I heard another uttering this cheerful strain:

In the city of Vancouver a rooster uttered his clear, long crow in three notes, as follows:

There was sometimes a slight falling in pitch at the close of the long final note.

At Sidcup, in November 1895, a fowl habitually drawled the following:

A cock-crow is usually of this character:

Here are two strains from the limited repertoire of the coal titmouse (heard in April):

This is a strain which I have heard sung from time to time by the great titmouse:

The "G's" in this were rather flat. I did not think it worth while to record many times the intervals sung by the chaffinch. Its song usually runs thus:

But on 26th May 1892, I heard at Brunswick Square, Gloucester, a chaffinch singing most unusual intervals; these:

This phrase was repeated many times.

The music of the robin (I mean in its original whistled notes) is sung so rapidly that it is very difficult to follow. The following are three of the most simple and distinct phrases which I have recorded as sung by this bird:

Of the musical intervals in alarm-cries I have recorded none but a few varieties of the blackbird's alarm. The following are three "alarms" which were heard at places about a mile apart:

The music of the hedge-accentor, like that of the robin, is rapid and difficult to follow. The three strains next recorded give a general idea of the song:

The blackcap is another rapid singer. The first

six strains of the following were sung by blackcaps at Stroud ; the rest at Leigh Woods, Clifton :

The notes of the mistle-thrush are easily recorded, because the bird is so feeble in originating, that it often reproduces the same strain, almost without variation, during several successive minutes. The following was sung by one in Cirencester Park on 21st May 1889:

At a spot a mile from Stroud, on 17th February 1889, another mistle-thrush sang :

Two others of this species had the following phrases respectively :

The American robin, which in form closely resembles the female blackbird, and which I have heard on many occasions, is even a poorer singer than the mistle-thrush, for its strains usually contain only two distinct notes, repeated many times in succession. The following are seven of its strains, heard in or near Vancouver:

I have always been much struck with the correctness of the time-accent observable in the music of blackbirds, thrushes, and skylarks. I have endeavoured to reproduce this, where possible, in the Appendix, by dividing the strains or songs of different individuals by means of bars. The songs were heard between March and August 1888, and the earlier records in relation to each species appear first. Many of them were repeated several times in succession. Double bars are placed between each strain. The music should be played (or whistled) in quicker time than that indicated in the stave.

CONCLUSION

IN the foregoing pages I have endeavoured to make some advance in the study of a highly interesting subject, which has not hitherto received any systematic treatment from ornithologists. Science has devoted attention too exclusively to the physical features of animals, and has neglected the study of their habits. Yet aberrant habit may be termed the parent of physical development. It suggests the future, as structure recalls the past. It also indicates thought and originality of purpose. In this country we may be said to live in the midst of birds; but, although the geographical distribution of these airy beings has been carefully recorded in all parts of the United Kingdom, their manners and various cries have hardly been noticed. Yet, of all pursuits in the realm of natural science, the study of these subjects may be considered the easiest to follow. The student needs only to be still and silent, and to look and listen to the wild-birds, which, as soon as they lose their fear, will impart to him some knowledge of their family secrets. He requires neither gun nor bag for carrying homeward any slain or captives: a field-glass, and a piece of waterproof cloth to sit upon, are all his equipment. The fields, trees, and hedges amongst which he moves will supply abundant incidents for investigation, and will amuse him with many a pretty scene from the infinite drama of nature. The reader who will take advantage of the opportunities thus afforded for observation will, I feel sure, arrive at the conviction that the evidence here adduced has been correctly stated, and that my conclusions are thereby justified.

APPENDIX

MUSIC OF THE BLACKBIRD

N.B.—*Large numbers distinguish the singers, double bars the strains.*

MUSIC OF THE THRUSH

Music of the Skylark

BIBLIOGRAPHY OF THE SUBJECT

THE following are the chief publications which contain information on bird-song. The authors are mentioned alphabetically :—

BARRINGTON, D. Experiments and Observations on the Singing of Birds (Roy. Soc. Phil. Trans., vol. lxiii., 1773, pp. 249-291).
BECHSTEIN, J. M. Natural History of Cage Birds.
BELON. Oiseaux. Paris, 1555.
BREE. Birds of Europe, not found in Britain.
CHAMBERS'S JOURNAL, vol. ix. No. 423.
CHENEY, SIMEON PEASE. Wood Notes Wild, Notations of Bird Music. Boston, 1892.
COLEMAN, A. P. Music in Nature (*Nature*, vol. xxxvi. p. 605).
DARWIN, CHARLES. Descent of Man.
   ,,   ,,   Expression of the Emotions.
FIELD, THE. Published weekly, London.
FOWLER, W. WARDE. Summer Studies of Birds and Books, 1895.
GARDINER, W. The Music of Nature. London, 1832.
GOULD, JOHN. Handbook of the Birds of Australia.
HARTING, J. E. Birds of Middlesex. London, 1866.
   ,,   ,,   Ornithology of Shakespeare. London, 1871.
   ,,   ,,   Our Summer Migrants.
HUDSON, W. H. Birds in a Village.
   ,,   ,,   Idle Days in Patagonia.
   ,,   ,,   The Naturalist in La Plata.
JESSE, ED. Gleanings in Natural History. London.
KIRCHER, A. Musurgia Universalis, lib. i. Romæ, 1590.
LAYARD. Birds of South Africa.
LESCUYER, F. Langage et Chant des Oiseaux. Paris, 1878.

LIBRARY OF ENTERTAINING KNOWLEDGE. Domestic Habits of Birds. London, 1833.
MACGILLIVRAY, W. History of British Birds.
MAGAZINE OF NATURAL HISTORY, vol. i. p. 346; vol. ii. p. 113.
MONTAGUE, COL. G. Ornithological Dictionary.
MONTBEILLARD. Oiseaux.
SAXBY, DR. H. L. The Birds of Shetland.
SMEE, A. My Garden. London, 1872.
STERLAND. Birds of Sherwood Forest.
SYME. British Song-birds.
WEBER, DR. F. On Melody in Speech (*Longman's Magazine*, vol. ix. 1877, pp. 399-411).
WHITE, GILBERT. Natural History of Selborne.
WILSON and BUONAPARTE. American Ornithology.
WITCHELL, C. A. The Evolution of Bird-song (*Zoologist*, July and August 1890).
,, ,, Bird-song and its Scientific Value (Proc. Cotteswold Naturalists' Field Club, vol. x. part 3).
,, ,, The Voice-Languages of Birds (Proc. Cheltenham Nat. Soc., 1892).
,, ,, The Scientific Value of Bird-song (*The Field*, 27th April 1895).
ZOOLOGIST, THE. A monthly publication, London.

INDEX

ALARM-NOTES, 22
  discrimination in, 27-30
  elaboration of, 23
  inducing silence, 31
  inheritance of, 30
  of blackbird, 24
  of young birds, 25
  origin of, 22
  relation to enemies, 26
  repetition of, 23
  variation of, 26, 236
American bluebird, 102
American robin, 97, 101, 104
Angola finch, 132
Archetypal cries, 137
Argentine blackbird, 111
Australian robins, 111

BECHSTEIN on mimicry, 192, 218
Bird-songs, defined, 10
Blackbird, 56, 65, 67, 73, 80, 95, 97, 166, 224
Blackbirds and dog, 29
Blackcap, 67, 68, 73, 150
Black-headed bunting (Am.), 122
Black redstart, 105
Blue jay (Am.), 94
Blue titmouse, 53, 55
Brambling, 126
Broad-billed flycatcher (S. Af.), 191
Brown wren, 51, 55, 73
Bullfinch, 34, 50, 56, 61, 172
Buntings and finches, 122

CALL-NOTE, 40
  abandoned, 47
  in song, 53
  repeated for song, 50

Call-note, repetition of, 49
Call-notes of allies alike, 68
Canary, 34, 40, 163, 172
Catbird (Am.), 165
Chaffinch, 42, 43, 55, 67, 147, 150, 225
Chaffinchs' battle-cry, 43
  call, 42
Chat thrush (S. Af.), 104
Chickadee (Am.), 135
Chicks, 86
Chiffchaff, 61
Cirl-bunting, 56, 68, 120, 187
Coal tit, 55
Cock of the rock (S. Am.), 67
Cock's crow, 26, 37, 51, 56, 64
Colymbidæ, 46
Combat, influence of, 33
  song in, 37
  wren's song in, 39
Common bunting, 55
Conclusion, 239
Confinement, effect of, 75
Creeper, 50
Crested screamer (S. Am.), 69, 81, 85
Cries, characters of, 47
  persistence of, 46
Crow of British Columbia, 93
Crows, 95, 163, 164, 175
Cuckoo, 89, 134, 143, 154, 155
Cushat, 51

DANCING, 66
Dartford warbler, 113
Dendrocolaptidæ, 60
Dog, use of, 99
Doves, 134
Ducks, 86, 89, 182

EAGLE owl (S. Af.), 188
Emberizidæ, 120
Environment, effects of, 177, 199, 227

FAMILY cries, 90
Fear, influence of, 45, 76
Females singing, 56
Ferruginous thrush (Am.), 111
Fieldfare, 96
Field-sparrow (Am.), 186
Finches and buntings, 122
First cries, 7
Fog, effect of, 64, 107
Fowl, 86, 89, 134
Frogs, 19
Frost, effect of, 194

GALLINACEOUS birds, 35
Garden warbler, 109
General and special types, 92
Golden-crested wren, 50, 67, 222
Golden plover, 87, 89
Goldfinch, 125, 126, 128, 151, 171
Goldfinch (Am.), 83, 132
Goose, 86
Grasshopper-warbler, 55, 56, 60, 68, 186
Greater nightingale, 104
Great tit, 55, 92
Green bulbul (India), 175
Greenfinch, 36, 50, 68, 80, 125-128, 169, 173
Green woodpecker, 67
Grunts (first cries), 14, 15

HAWFINCH, 172
Hedge-sparrow, 50, 73, 147
Heredity, 86-139
  and imitation, 90
  in Anatidæ, 138
  in blackbird and robin, 97
  in blackbird and thrush, 110
  in blackcap and garden-warbler, 109
  in buntings and finches, 123
  in buntings and pipits, 120
  in chaffinch and brambling, 131

Heredity, in chaffinch and house-sparrow, 129
  in chick, 86
  in Corvidæ, 93
  in Corvidæ and Turdidæ, 95
  in cuckoos, 134
  in finches, 125
  influence of, 86
  in greenfinch and brambling, 126
  in greenfinch and canary, 125
  in lapwing, 88
  in larks and pipits, 118, 119
  in mistle-thrush and American robin, 107, 108
  in mistle-thrush and ouzel, 108
  in mistle-thrush and redwing, 138
  in nightingale and lesser whitethroat, 107
  in nightingale and sedge-warbler, 107
  in partridge, pheasant, and fowl, 134
  in pheasant, etc., 89
  in pigeons and doves, 134
  in pipits, 102
  in raptorial birds, 134
  in redstart and flycatcher, 99, 104
  in redstart and nightingale, 105
  in redstart and robin, 97, 98
  in redstart and warblers, 99, 104
  in redstart and whitethroat, 105
  in rhea, 88
  in robin and blackbird, 104
  in robin and hedge-sparrow, 114
  in robin and nightingale, 104
  in sparrows, notes of, 129-131
  in thrushes and warblers, 95, 100-104, 117
  in tinamu, 88
  in Turdidæ, 95
  in wagtails and pipits, 117, 118
  in warblers compared, 113
  in willow-warbler and yellow-bird (Am), 106
  in woodpeckers, 134, 143
Hissing, 15-17
Homing pigeon, 66
Hooded crow, 93, 146
House-sparrow, 61, 153, 154, 160, 169

INDEX

IMITATION, 4, 159-229
and heredity, 162
filial, 160
in blackbird, 224
in bullfinch, 172
in canaries, 163, 172
in captive birds, 162
in catbird (Am.), 165
in chaffinch, 225
in crows, 163
in dog, 160
in finches, 169
in fish, 161
influence of, 159
in foreign birds, 174
in goldcrest, 222
in goldfinch, 171
in green bulbul (India), 175
in greenfinch, 169, 172
in hawfinch, 172
in house-sparrow, 169
in hybrid finches, 172
in jay, 164, 165
in linnet, 170, 172
in marsh-warbler, 220
in menuras (Australia), 174, 175
in minah (India), 175
in mistle-thrush, 165
in mocking-bird (Am.), 165
in nightingale, 218
in nutcracker, 164
in nuthatch, 223
in redstart, 192, 218
in reed-bunting, 223
in reed-warbler, 221
in robin, 195, 206-208
in sedge-warbler, 197, 214
in shrikes, 173, 174
in skylark, 196, 208
in starling, 196, 211, 233
in stonechat, 224
in thrush, 194, 200
in thrushes, 165
in warblers, 167
in wheatear, 168, 222
in whinchat, 168
in whitethroats, 223
observations on, 175-181
of birds, 190
of elements, 181

Imitation of insects, 186
of quadrupeds, 187
survivals of, 227
Indigo bird (Am.), 122
Instrumental music, 143
Introduction, 1

JACKAL buzzard (S. Af.), 188
Jackdaw, 55, 93

KESTREL, 49, 53, 79
Kingbird (Am.), 37
Kingfisher, 155

LABORIOUS birds, 79
Landrail, 189
Lapland finch, 132
Lapwing, 87, 89
Lark, mimicry of, 196, 208
music of, 245
Larks and pipits, 118
Leisure, necessary, 77
Lesser spotted woodpecker, 50
Linnet, 68, 83, 152, 170, 172

MARSH-TIT, 42
Marsh-warbler, 220
Menuras (Australia), 69, 174, 175
Minah (India), 175
Mistle-thrush, 64, 95, 96, 110, 165
Mocking-bird (Am.), 165
Music, antiquity of, 232
of American robin, 238
of bird-song, 230
of blackbird, 241
of blackbird's alarm, 236
of blackcap, 237
of chaffinch, 235
of coal titmouse, 235
of cock-crows, 234
of hedge-sparrow, 236
of mistle-thrush, 237
of robin, 235
of skylark, 238, 245
of starling and parrot, 233
of streams, 184
of thrush, 231, 244
pitch not noticed, 233

NESTLING, education of, 171
Newts, 17
Newts and snakes, 18
Nightingale, 1, 52, 55, 62, 67, 109, 144, 166, 218
Nightjar, 68, 186
Northern diver, 63
Noticeable incidents of song, 57
Nutcracker, 94, 164
Nuthatch, long note of, 54

OLIVACEOUS thrush (S. Af.), 96
Ortolan bunting, 121
Osprey, 63
Ostrich, 188
Oven bird (S. Am.), 60, 61, 89

PARTRIDGE, 86, 89, 134
Peregrine falcon, 17
Perpetuation of cries, 47
Persistence of cries, 135
Pheasant, 86, 89, 134
Phrase, defined, 10
Picus pileatus (Am.), 63
Pied wagtail, 51
Pine finch (Am.), 132
Pine grosbeak, 50
Pinnated grouse (Am.), 37
Pipits, 117, 118, 120
Poetical writers, 9
Polygamous species, 35

RAPTORES, 46
Rasorial birds, 79
Recording, method of, 4
Red-headed woodpecker (Am.), 187
Redpoll, 68
Redstart, 56, 82, 97, 105, 192, 218
Redwing, 94
Red-winged starling (Am.), 83
Reed-bunting song, 122, 123, 223
Reed-warbler, 68, 221
Repetition of intervals, 57
Rice-bunting (Am.), 123
Rice troupial (Am.), 83
Ringed plover, 50
Ring ouzel, 96
Robin, 34, 51, 61, 63, 65, 77, 97, 103, 152, 166, 195, 206-208

Rook, 49, 93, 146
Rupicola (S. Am.), 67

SCARLET bullfinch, 191
Sedge-warbler, 68, 80, 197, 214-217
Shrikes, 27, 173, 174, 191
Shrike thrush (Australia), 111
Siskin, 68
Skylark, 34, 52, 54, 56, 61, 82, 120, 196, 208-211
Snake, 189
Snow-bunting, 37, 120, 122
Song, accent in, 73, 238
Song apparatus in crows, etc., 70, 72
Song-birds and sober hues, 73
Song, by adults, 61
  by males, 59
  effect of fog, 64
  flight in, 80
  in chorus, 83
  increased vehemence, 67
  in rain, 63
  laborious birds and, 79
  leisure for, 75
  morning and evening, 61
  only in small birds, 68
  varied at end, 68
  ventriloquism, 82
Songs, rise in pitch, 67
  the simplest, 49
Squirrel, 189
Starling, 53, 56, 67, 77, 83, 94, 152, 196, 211-214
Stock-dove, 51, 53, 68
Stonechat, 55, 224
Strain, defined, 11
Swallow, 29, 55
Swallow's alarm, 29
Swan, 89, 155
Swift, 83, 182, 183
Surroundings, pleasure in, 65

THRUSH, 2, 61, 65, 73, 95, 96, 110, 194, 200-206
Thrushes and warblers, 95, 117
Tinamu, 54, 88, 89
Titmice, 55, 134
Towhe bunting (Am.), 122

INDEX

Tree-creepers, 60
Tree-pipit, 55, 82
Turkey, 86, 89
Tyrant flycatcher (Am.), 60

VANCOUVER wren, 135
Variation, 140
  absence of, 144
  at end of song, 156
  in blackbird's alarm, 236
  in blackbird's song, 108, 148
  in chaffinch, 148, 235
  in cock-crow, 234
  in robin's alarm, 157
  in speed, 157
  instances of, 146
  local, 146, 149
  progressive, 36, 148
  transient, 153
Voice, accidental, 17-19

Voice, Darwin's theory of, 13
  origin of, 12

WARBLERS and thrushes, 95, 117
Wheatear, 168, 222
Whinchat, 168
Whitethroats, 68, 105, 107, 112, 113, 223
Willow-warbler, 52, 56, 61, 65, 67, 147, 149, 150, 151
Wings, moved in song, 56
Woodhewers (S. Am.), 60
Woodlark, 65
Woodpeckers, 46
Wood-wren, 56, 67, 80, 145

YELLOW bunting, 51, 55, 67, 187
Young birds, cries of, 44

ZEBRA finch, 53

THE END

Printed by R. & R. CLARK, LIMITED, *Edinburgh*

In 4 parts, Demy 8vo, Boards, Leather Back.   Price 7s. 6d. net.

A DICTIONARY OF BIRDS

BY

ALFRED NEWTON

ASSISTED BY

HANS GADOW

WITH CONTRIBUTIONS BY

RICHARD LYDEKKER, B.A., F.G.S., CHARLES S. ROY, M.A., F.R.S.

AND

ROBERT W. SHUFELDT, M.D., Late United States Army.

TO BE COMPLETED IN FOUR PARTS.

Parts 1, 2, and 3 now ready.

Price 7s. 6d. net, each.

"The appearance of a general work on birds by an ornithologist of the long experience of Professor Newton may well be regarded as marking an epoch in the science of which it treats. It is the best book of its kind which has yet appeared."—*Natural Science.*

"Will probably be the most useful and accurate compendium of the subject in any language."—*Nature Notes.*

"The work was much wanted, and contains quite enough to make it of very high value."—*Land and Water.*

"Promises to prove an extremely useful work."—*Saturday Review.*

"Professor Newton has produced the *magnum opus* of his life, and when completed the work will not only be a lasting monument of his great researches and close ornithological observations, but an invaluable and safe guide for all students of one of the most delightful branches of human investigation."—*Yorkshire Post.*

A. & C. BLACK, SOHO SQUARE, LONDON, W.

In One Volume, Fcap. 4to, containing 22 page Illustrations, and 11 Diagrams in the text. Bound in Buckram, gilt top. Price 21s.

ARTISTIC & SCIENTIFIC TAXIDERMY AND MODELLING

A MANUAL OF INSTRUCTION IN THE METHODS OF PRESERVING AND REPRODUCING THE CORRECT FORM OF ALL NATURAL OBJECTS INCLUDING A CHAPTER ON THE MODELLING OF FOLIAGE

BY

MONTAGU BROWNE, F.G.S., F.Z.S., ETC.

CURATOR OF THE LEICESTER CORPORATION MUSEUM AND ART GALLERY
AUTHOR OF 'PRACTICAL TAXIDERMY,' 'THE VERTEBRATE ANIMALS OF LEICESTERSHIRE AND RUTLAND,' ETC.

With 22 Full-page Illustrations and 11 Illustrations in Text

A. & C. BLACK, SOHO SQUARE, LONDON, W.

Mitchell, C
MN
AUG 16 1
FEB 2 0 1

www.ingramcontent.com/pod-product-compliance
Lightning Source LLC
Chambersburg PA
CBHW032000230426
43672CB00010B/2224